McGraw-Hill's
GED
Science

WORKBOOK

*The Most Thorough Practice for
the GED Science Test*

Robert Mitchell

McGraw·Hill

New York   Chicago   San Francisco   Lisbon   London   Madrid   Mexico City
Milan   New Delhi   San Juan   Seoul   Singapore   Sydney   Toronto

1 2 3 4 5 6 7 8 9 0   QPD/QPD   1 0 9 8 7 6 5 4 3 2

ISBN 0-07-140705-7

McGraw-Hill books are available at special quantity discounts to use as premiums and sales promotions, or for use in corporate training programs. For more information, please write to the Director of Special Sales, Professional Publishing, McGraw-Hill, Two Penn Plaza, New York, NY 10121-2298. Or contact your local bookstore.

This book is printed on acid-free paper.

Table of Contents

Acknowledgments

Drawing of Pascal's calculator on page 21: St. Andrews University, Scotland

Illustration of Benjamin Franklin's electricity experiment on page 22: © Bettmann/CORBIS

Photo of Edison's phonograph on page 22: © Bettmann/CORBIS

Photo of mushroom-shaped rock on page 82: © Joseph Sohm; ChromoSohm Inc./CORBIS

Photo of anemometer on page 82: © Ecoscene/CORBIS

Photo of Hubble Space Telescope on page 84: © NASA

Photo of Moon on page 85: © NASA

Photo of lunar landing on page 85: © NASA

Photo sequence of solar flare on page 86: © NASA

Photo of epithelial tissue on page 100: © Jim Zuckerman/CORBIS

Photo of nervous tissue on page 100: © Lester V. Bergman/CORBIS

Photo of muscle tissue on page 100: © Lester V. Bergman/CORBIS

Photo of connective tissue on page 100: © Lester V. Bergman/CORBIS

Infared image of hurricane on page 102: © NOAA

Photo of seismograph reading on page 103: © U.S. Geological Survey

Introduction

McGraw-Hill's GED Science Workbook will help you study for the GED Science Test. There are four main sections in this exercise book: Themes in Science, Life Science, Physical Science, and Earth and Space Science. Each section gives you additional practice in an area covered in *McGraw-Hill's GED Science* and in the science portion of *McGraw-Hill's GED*. The page numbers at the beginning of each chapter in the exercise book will refer you to the appropriate pages in either text.

This exercise book also contains a full-length **Practice Test**. This test is very similar in length and format to the actual GED Science Test.

Overview of the GED Science Test

The GED Science Test consists of multiple-choice questions. These questions require you to know some basic science concepts and to be able to think about these concepts. For most questions, you will have to read a passage or look at an illustration and answer questions based on it. In some cases you will have to rely on prior scientific knowledge in order to answer the question correctly.

There are 50 multiple-choice questions, and you will be given 80 minutes to complete the test. About half of the questions will be based on diagrams, charts, or graphs. The other half will be based on short reading passages.

Content Areas

The GED Science Test is based on the National Science Education Standards of the National Academy of Sciences. In accordance with these standards, emphasis is placed on both fundamental content areas and interdisciplinary themes—themes that are common to all areas of science.

The passages and graphics on the GED Science Test are taken primarily from the following content areas:

Content Area	Percentage of Test
Life Science *(Plant and Animal Science; Human Biology)*	45%
Physical Science *(Chemistry; Physics)*	35%
Earth and Space Science	20%

While 60 percent of the questions will focus solely on these content areas, keep in mind that a given question may draw from more than one of these topics. It's hard to discuss science without touching on a number of topics. For example, a question on air pollution may draw from material covered in both biology and chemistry.

The remaining 40 percent of the questions on the test will deal with the following interdisciplinary themes as they relate to the content areas of science:

Interdisciplinary Themes

- Unifying Concepts and Processes
- Science as Inquiry
- Science and Technology
- Science in Personal and Social Perspectives
- History and Nature of Science

At the end of the book is a complete **Answer Key** that tells you the reasoning behind each correct answer choice. Be sure to check your answers and to read the explanations. This will help you improve your skill in answering multiple-choice questions.

To determine whether or not you are ready to take the real GED Science Test, we recommend that you take the **Practice Test** at the end of this book. The evaluation chart that follows the test will help you determine the areas in which you may need additional practice.

If you need extra preparation, the **Science Almanac** on pages 117–121 provides lists and tables of useful scientific information along with links to Web sites containing additional instruction and practice.

Concepts and Processes in Science

GED Science pages 27–48

Directions: Choose the <u>one best answer</u> to each question.

Questions 1–3 refer to the following passage.

Plants are often classified according to their characteristic features and life cycle. Five types of plants are described below.

Annuals germinate from seeds, produce flowers, and die in one growing season.

Vegetable biennials are vegetables that have a two-year life cycle. The seeds germinate the first year and produce roots, a short stem, and leaves. In the second year the stem grows, and flowers, fruits, and seeds are produced. Vegetable biennials are usually harvested during their first year of growth.

Flower biennials have the same characteristics as vegetable biennials. The difference between the two types of biennials is that flower biennials are known for their flowers and are not harvested as food.

Herbaceous perennials are plants that live for more than two growing seasons. The roots of these perennials grow slowly during the winter and produce new shoots each spring. Herbaceous perennials are foliagelike (soft and leafy) and do not contain wood.

Woody perennials also live for more than two growing seasons. But unlike herbaceous perennials, woody perennials stop growing during the winter and resume growth in the spring. Woody perennials are characterized by stiff wooden trunks and branches.

1. The seeds of a pea plant must be planted each year to start new plants. The rest of the plant dies at the end of the growing season.

   Based on the information above, how would you classify the pea plant?

   (1) annual
   (2) vegetable biennial
   (3) flower biennial
   (4) herbaceous perennial
   (5) woody perennial

2. Carrots, unlike pea plants, have a two-year life cycle. However, carrots are usually harvested for consumption during their first year of growth.

   Based on the information above, how would you classify carrots?

   (1) annuals
   (2) vegetable biennials
   (3) flower biennials
   (4) herbaceous perennials
   (5) woody perennials

3. Peonies are a favorite type of yard flower that resurface year after year after they are planted. At the end of each growing season, the aboveground part of the plant dies.

   Based on the information above, how would you classify peonies?

   (1) annuals
   (2) vegetable biennials
   (3) flower biennials
   (4) herbaceous perennials
   (5) woody perennials

Questions 4–7 refer to the illustration below.

GLUCOSE FEEDBACK CONTROL

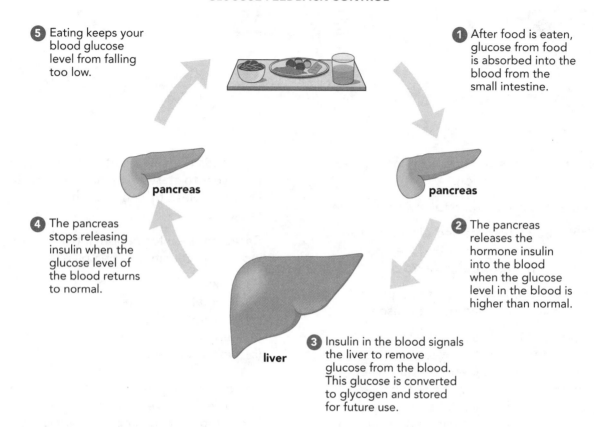

5 Eating keeps your blood glucose level from falling too low.

1 After food is eaten, glucose from food is absorbed into the blood from the small intestine.

pancreas

pancreas

4 The pancreas stops releasing insulin when the glucose level of the blood returns to normal.

2 The pancreas releases the hormone insulin into the blood when the glucose level in the blood is higher than normal.

liver

3 Insulin in the blood signals the liver to remove glucose from the blood. This glucose is converted to glycogen and stored for future use.

4. What is the name of the hormone that regulates the amount of blood sugar (glucose)?

   (1) estrogen
   (2) pancreas
   (3) insulin
   (4) glycogen
   (5) liver bile

5. Which of the following can be inferred from the illustration above?

   (1) The human body functions most efficiently with high levels of blood sugar.
   (2) The human body functions most efficiently with a limited amount of blood sugar.
   (3) The human body functions most efficiently with no blood sugar.
   (4) The human body uses blood sugar to control insulin level in the blood.
   (5) Excess blood sugar is removed from the blood and stored in the pancreas.

6. The human body's sugar-regulating mechanism is an example of what general scientific process?

   (1) photosynthesis
   (2) respiration
   (3) metamorphosis
   (4) evolution
   (5) homeostasis

7. Which household device plays a regulatory role that is analogous (similar) to the role played by the pancreas in the human body?

   (1) thermostat
   (2) smoke detector
   (3) doorbell
   (4) microwave oven
   (5) shower

Question 8 refers to the following drawing.

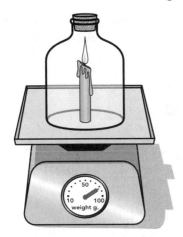

8. Which of the following will <u>not</u> change as the candle burns?

(1) the height of the flame
(2) the amount of wax in the candle
(3) the length of the wick
(4) the amount of oxygen gas in the jar
(5) the weight shown on the scale

Question 9 refers to the following information.

Candice flips a penny 19 times in a row. She records the results of her flips in a tally chart.

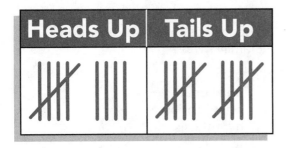

Heads Up	Tails Up

9. Suppose Candice flips the penny one more time. Which statement below is true?

(1) The twentieth flip is more likely to be heads up than tails up.
(2) The twentieth flip is more likely to be tails up than heads up.
(3) The twentieth flip is equally likely to be heads up or tails up.
(4) The twentieth flip is not likely to be either heads up or tails up.
(5) The twentieth flip is less likely to be tails up than heads up.

Questions 10 and 11 refer to the following diagram.

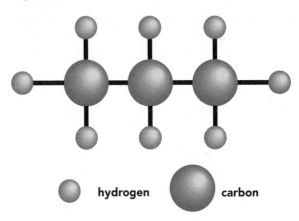

hydrogen   carbon

10. What does the model represent?

(1) a chemical reaction
(2) an atom
(3) an atomic nucleus
(4) a molecule
(5) an electric current

11. Which chemical formula describes this model?

(1) 3C8H
(2) C_3H_8
(3) 3C + 8H
(4) $C_3 + H_8$
(5) 11CH

12. What does not change during a chemical reaction?

(1) the total number of atoms present
(2) the total number of molecules present
(3) the types of molecules present
(4) the volume taken up by the substances that are reacting
(5) the amount of heat energy present

Questions 13 and 14 refer to the following passage and illustration.

Things tend to happen in familiar ways. The illustration below shows four steps of a short experiment a student performed with a group of marbles. The steps, labeled A through D, are not shown in the order in which they actually occurred.

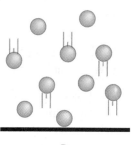

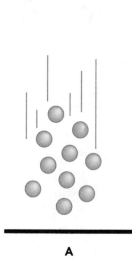

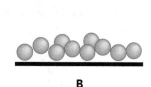

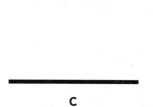

A   B   C   D

13. In what sequence did the steps shown above actually occur?

(1) A, C, B, D
(2) C, B, A, D
(3) A, B, D, C
(4) C, A, D, B
(5) B, D, A, C

14. The student performs the experiment three more times. In each experiment, the same sequence of steps occurs.

What word is used to describe the tendency of an event to happen in a familiar way?

(1) order
(2) equilibrium
(3) classification
(4) hypothesis
(5) form

15. With which other animals is a sea horse most likely to be scientifically classified?

(1) sparrows and condors
(2) sharks and tuna
(3) mosquitoes and grasshoppers
(4) sea turtles and lizards
(5) frogs and salamanders

16. What characteristic do reptiles and mammals have in common?

(1) the laying of an amniotic egg
(2) body hair
(3) a backbone enclosing a spinal cord
(4) mammary glands
(5) being cold-blooded

Answers are on page 107.

Comprehending and Applying Science

GED Science pages 49–66

Directions: Choose the <u>one best answer</u> to each question.

Question 1 refers to the following illustration.

1. What property of sunlight is being shown in the illustration above?

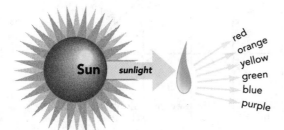

   (1) Sunlight creates raindrops.
   (2) Sunlight warms raindrops.
   (3) Sunlight cannot pass through water.
   (4) Sunlight is made of a spectrum of colors.
   (5) Depending on the time of day, a raindrop may be one of many colors.

2. The law of conservation of matter states that matter is neither created nor destroyed during a chemical reaction. The number of atoms of each element must be the same on both sides of a chemical equation.

   Which equation shows one molecule of propane gas (C_3H_8) combining with five molecules of oxygen (O_2) to create several molecules of carbon dioxide (CO_2) and water (H_2O)?

   (1) $C_3H_8 + 5O_2 \longrightarrow 3CO_2 + 2H_2O$
   (2) $C_3H_8 + 5O_2 \longrightarrow 3CO_2 + 4H_2O$
   (3) $C_3H_8 + 5O_2 \longrightarrow 4CO_2 + 2H_2O$
   (4) $C_3H_8 + 5O_2 \longrightarrow 4CO_2 + 4H_2O$
   (5) $C_3H_8 + 5O_2 \longrightarrow 4CO_2 + 5H_2O$

3. In any ecosystem organisms compete for limited resources. Competition is most severe between members of the same species. However, it also occurs between organisms that are similar but not identical.

   Between which pair of animals is competition <u>least</u> likely to occur?

   (1) a whale and a dolphin
   (2) two wolves
   (3) a squirrel and a chipmunk
   (4) a deer and a blackbird
   (5) a sparrow and a robin

Question 4 refers to the illustration below.

4. The embryos of the vertebrates above display a surprising similarity. What is the most reasonable interpretation of this evidence?

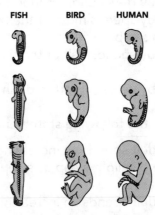

   (1) Vertebrates develop into similar adults.
   (2) Vertebrates have similar nutritional needs.
   (3) Vertebrates develop similar brains.
   (4) Vertebrates share a common ancestor.
   (5) All vertebrates have four legs.

Questions 5–7 refer to the following passage and graph.

By monitoring electrical impulses given off by the brain, scientists have discovered that people alternate between periods of light sleep and periods of deep sleep. The two sleep states seem to play different roles for the health of the body and the mind. During light sleep the body is restless and moves around on the bed; the mind is very active, and dreaming takes place. Because rapid eye movement takes place during light sleep, light sleep is often referred to as REM sleep.

During deep sleep, the body is much less active, and little, if any, dreaming occurs. Deep sleep is often referred to as non-REM (NREM) sleep.

Infants alternate equally between these two sleep states. Adults, on the other hand, spend most of their time in the deep-sleep state.

The typical sleep pattern of an adult is illustrated below.

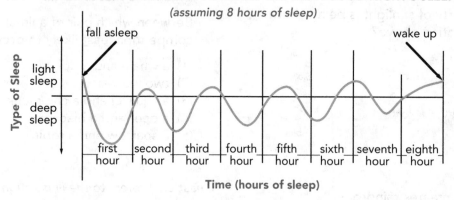

TYPICAL SLEEP PATTERN OF AN ADULT HAVING A GOOD NIGHT'S SLEEP
(assuming 8 hours of sleep)

5. What is one difference between the way a baby sleeps and the way an adult sleeps?

   (1) A baby cries often during sleep.
   (2) A baby gets hungry during sleep.
   (3) A baby spends most of the time in deep sleep.
   (4) A baby spends half of the time in light sleep.
   (5) A baby wakes up often during sleep.

6. Which of the following statements is true?

   (1) An adult spends an increasing amount of time in light sleep as sleep time increases.
   (2) An adult spends an increasing amount of time in deep sleep as sleep time increases.
   (3) An adult spends most of the first sleeping hour in light sleep.
   (4) An adult sleeps more deeply during the later hours of sleep than during the earlier hours.
   (5) An adult wakes up during a deep-sleep period.

7. Which of the following is a conclusion you can draw from the passage and graph?

   (1) An adult is likely to remember a dream if he or she wakes up after four hours of sleep.
   (2) Adults spend more time sleeping than children do.
   (3) An adult who kicks off the covers during sleep spends more time in light sleep than in deep sleep.
   (4) Adults who sleep less than eight hours do not get enough sleep.
   (5) An adult does the most dreaming during the final hour of sleep.

Questions 8–10 refer to the graph below.

30-YEAR STUDY OF DEER POPULATIONS IN SHERMAN FOREST

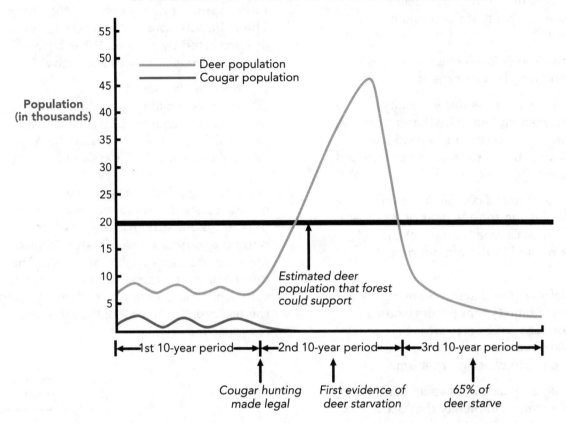

8. Which phrase best describes what happened to the deer population after the decline of the cougar population in Sherman Forest?

   (1) It increased continuously.
   (2) It increased to above normal level for several years and then decreased to below normal level.
   (3) It decreased to below normal level for several years and then increased to above normal level.
   (4) It decreased to its lowest level.
   (5) It reached its maximum stable level at 20,000 deer.

9. What most likely led to a rapid decline of the deer population?

   (1) inadequate water supply
   (2) lack of sufficient living space
   (3) inadequate food supply
   (4) increase in fatal diseases
   (5) increase in level of air pollution

10. A *predator* is an animal that kills and eats other animals (called *prey*). Cougars, for example, are predators, and deer are their prey. With these definitions in mind, which of the following generalizations is best supported by information given in the first ten-year segment of the graph?

    (1) Predators and prey search for food during different times of the day.
    (2) Predators and prey search for food during the same time of the day.
    (3) An increase in a predator population always follows a decrease in a prey population.
    (4) A predator population reaches its peak after a prey population reaches its peak.
    (5) A predator population reaches its peak before a prey population reaches its peak.

Questions 11–13 refer to the following passage.

Immunity is freedom from catching a certain disease. Five types of immunity are listed below.

Inherited immunity—immunity that is inherited and may be permanent

Naturally acquired active immunity—an often permanent immunity that occurs when a person's body naturally produces antibodies after being exposed to or infected by a disease

Naturally acquired passive immunity—an immunity lasting for one year or less that occurs in a fetus or small infant because of antibodies passed to the offspring by the mother

Artificially acquired active immunity—a long-term immunity that occurs when a person produces antibodies after being injected with a vaccine containing dead or weakened disease-causing organisms

Artificially acquired passive immunity—usually a short-term immunity that occurs when antibodies produced in an animal are injected into a person

11. Studies show that children who were breastfed as babies are less likely to get certain diseases than children fed with artificial baby formula. Doctors suspect that mother's milk provides antibodies that are not present in formula. Assuming that this is true, what type of immunity is provided by mother's milk?

    (1) inherited immunity
    (2) naturally acquired active immunity
    (3) naturally acquired passive immunity
    (4) artificially acquired active immunity
    (5) artificially acquired passive immunity

12. For protection against tetanus, an often fatal disease caused by bacteria, doctors recommend that children be given a tetanus shot at about two months of age. The antibodies present in a tetanus shot are produced by a horse. What type of immunity does a tetanus shot provide?

    (1) inherited immunity
    (2) naturally acquired active immunity
    (3) naturally acquired passive immunity
    (4) artificially acquired active immunity
    (5) artificially acquired passive immunity

13. The line graph below shows how the natural concentration of antibodies in a person's bloodstream changes when the body is exposed at two different times to the same disease-causing organism. The primary immune response is often not sufficient to keep the person from getting the disease, but the secondary immune response may be.

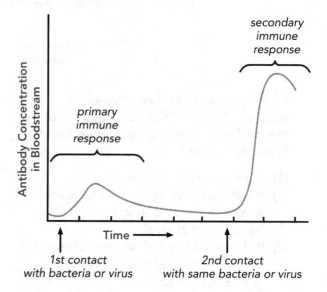

What type of immunity is represented on this graph?

    (1) inherited immunity
    (2) naturally acquired active immunity
    (3) naturally acquired passive immunity
    (4) artificially acquired active immunity
    (5) artificially acquired passive immunity

Answers are on page 107.

Analyzing and Evaluating Science

Directions: Choose the <u>one best answer</u> to each question.

Questions 1 and 2 refer to the following passage.

During an evening storm, Maria is sitting in her living room. A lamp is on, and her radio is playing. As the room gets colder, Maria plugs a space heater into a nearby wall socket. A few seconds later, the lamp goes out but the radio keeps playing. Maria notices that the space heater has also gone off.

Maria has the following thoughts:

A. The bulb in the lamp has burned out.
B. The radio is battery-powered.
C. The power has gone off.
D. I should see if the kitchen lights work.
E. The lamp and the heater are plugged into the same wall socket.

1. Which of Maria's thoughts are hypotheses that suggest reasons why the lamp stopped working?

   (1) Both A and B
   (2) Both A and C
   (3) Both B and D
   (4) A, C, and E
   (5) B, C, and D

2. Suppose Maria discovers that the lights in the kitchen are still working. What is the most reasonable conclusion she can draw, knowing that the lamp and heater in the living room do not work?

   (1) The kitchen and the living room are not on the same circuit.
   (2) Electric service has gone off.
   (3) The bulb in the living room lamp is burned out.
   (4) The space heater is broken.
   (5) The kitchen lights use less electricity than the living room lamp.

Question 3 refers to the illustration below.

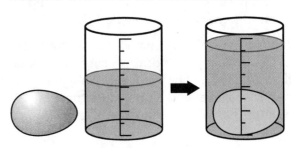

3. What can the experiment shown be used to measure?

   (1) the height of the egg
   (2) the volume of the egg
   (3) the weight of the egg
   (4) the water content of the egg
   (5) the surface area of the egg

Question 4 refers to the illustration below.

WHY THE SKY IS BLUE

air molecules

white sunlight

transmitted sunlight

scattered blue light

4. Why is the sky blue?

   (1) Sunlight has more blue light than any other color.
   (2) Gas molecules in air scatter blue light more than any other color.
   (3) Gas molecules in air absorb blue light more than any other color.
   (4) The ocean color causes the sky to appear blue.
   (5) Blue is the color of outer space.

Questions 5 and 6 refer to the following passage.

The boiling point of water depends on *air pressure*—the pressure on the top surface of the water. The boiling point decreases as the air pressure decreases. After water begins to boil, its temperature does not rise above the boiling point, even if more heat is added.

• Pure water boils at 212°F at sea level. At an altitude of 3,500 feet above sea level, where the air pressure is much lower, the boiling point of pure water is only 208°F.

• The boiling point of pure water on the floor of Death Valley, California, is slightly greater than 212°F. The floor of Death Valley is below sea level and air pressure there is greater than air pressure at sea level.

5. While vacationing in the mountains, Gretta cooked corn in boiling water for eight minutes. The next day at home on the coast, she again boiled corn for eight minutes. Compared to the corn in the mountains, how did the corn at home taste?

   (1) more cooked
   (2) less cooked
   (3) more salty
   (4) less salty
   (5) cooler

6. A pressure cooker is a covered container used for cooking food. As water boils inside, the top of the pressure cooker prevents the steam from escaping easily, thus increasing the pressure over the water. What is the main advantage of a pressure cooker compared to boiling in an open pan?

   (1) The boiling water can be reused.
   (2) Water needs less heat energy to be brought to a boil.
   (3) Water can boil is less time.
   (4) The boiling temperature of the water can be increased.
   (5) The boiling temperature of the water can be decreased.

Question 7 refers to the illustration below.

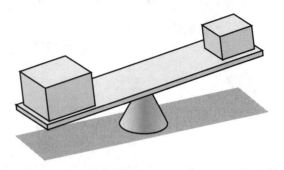

7. What is the device shown above capable of comparing?

   (1) the temperature of two objects
   (2) the dimensions of two objects
   (3) the surface areas of two objects
   (4) the volume of two objects
   (5) the weights of two objects

Question 8 refers to the following illustration.

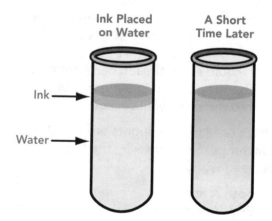

Ink Placed on Water   A Short Time Later

Ink →

Water →

8. Ink placed gently on water slowly begins to mix with the water in a process known as diffusion. What is most likely responsible for the process of diffusion?

   (1) random molecular motion
   (2) differences in molecular weight
   (3) gravitational force
   (4) electric current
   (5) static electricity

Questions 9–11 refer to the passage and diagram below.

If you stir a spoonful of sugar into a glass of water, the sugar quickly dissolves and disappears. You know this if you've ever made fresh lemonade, coffee, or any other drink that contains water and sugar.

One entire cup of sugar will dissolve in one cup of water. If you try to dissolve more than one cup, sugar will settle on the bottom. When the water contains all the sugar it can hold, the solution is described as saturated.

The drawing below shows what happened when Carlos poured sugar into a glass of apple cider. Carlos does not tell you how much sugar he added to the glass.

apple cider

sugar

9. Knowing that apple cider is mainly water, which of the following most likely explains why sugar is on the bottom of the glass?

   (1) Sugar does not dissolve in apple cider.
   (2) The sugar was placed in the glass before the apple cider was added.
   (3) The apple cider was placed in the glass before the sugar was added.
   (4) The apple cider in the glass is saturated with sugar.
   (5) Sugar will dissolve in apple cider only when the cider is heated.

10. Suppose the cider is already saturated with sugar. What feature of the solution will <u>not</u> change if more sugar is added to the glass?

    (1) the amount of sugar on the bottom of the glass
    (2) the weight of the glass and its contents
    (3) the height of the cider in the glass
    (4) the total amount of sugar in the glass
    (5) the color of the cider in the glass

11. Which of the following experiments would enable you to determine for sure whether or not sugar dissolves in apple cider?

    A. Add more sugar to the glass to see if the added sugar settles on the bottom.

    B. Add unsweetened apple cider to the glass to see if the amount of sugar on the bottom decreases.

    C. Carefully pour half of the apple cider out of the glass to see if more sugar appears on the bottom.

    (1) A only
    (2) B only
    (3) C only
    (4) Both A and C
    (5) Both B and C

Questions 12 and 13 refer to the passage and graph below.

The graphs below are called energy distribution charts. Each chart shows the average amount of light generated in each color band by the light source indicated: chart A represents natural outdoor light (sunlight on a clear day); chart B represents a typical fluorescent light.

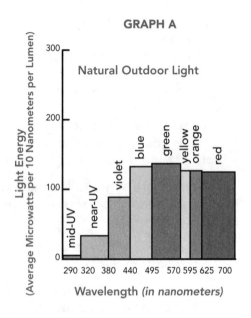

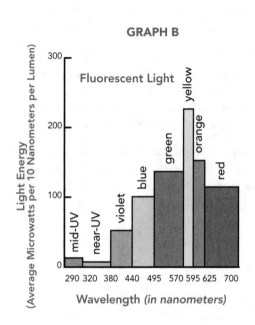

12. Which two of the following conditions are most likely to affect the energy distribution of sunlight that reaches your eyes?

   A. rain
   B. wind
   C. smog
   D. temperature

   (1) A and C
   (2) A and D
   (3) B and C
   (4) B and D
   (5) C and D

13. If a piece of paper appears bright white when viewed outside on a clear day, how would it appear when viewed in a room lit only by the fluorescent light represented by graph B?

   (1) bright white, the same as in outdoor light
   (2) slightly red
   (3) slightly blue
   (4) slightly yellow
   (5) slightly violet

   Answers are on page 108.

Science and Technology

GED Science pages 95–114

Directions: Choose the <u>one best answer</u> to each question.

Questions 1–3 refer to the passage below.

A microwave oven uses *microwaves* (low-energy light waves) to cook food. When the oven is running, electricity flows to the magnetron, a device that changes electrical energy into microwaves. The microwaves are given off by the antenna and are guided to the stirrer by a hollow tube known as a waveguide. The stirrer reflects the microwaves evenly throughout all parts of the oven's interior. Reflective surfaces on the microwave's walls and a reflective screen on the see-through door help ensure an even distribution of microwave energy throughout the oven.

Microwaves are absorbed by water, fats, sugars, and some other molecules in food. When these molecules are evenly dispersed in many foods, those foods will cook evenly and quickly from the inside, without heating the surrounding air. This is unlike a conventional oven, which heats food by hot air, cooking the outer layers of the food first, and heating all inner surfaces of the oven as well.

A microwave oven also differs from a conventional oven in that materials such as glass, paper, and most plastics can be safely used as cookware. Microwaves pass through these materials and are not absorbed.

1. What is the most likely reason that Juan's microwave oven cooked a turkey well done on one side and only partially done on the other?

   (1) The power cord was broken.
   (2) The oven door was open during cooking.
   (3) The magnetron did not work properly.
   (4) The antenna did not work properly.
   (5) The stirrer did not work properly.

2. While heating pizza on a paper plate in her microwave oven, Yoko noticed that the plate got hot. Why did this most likely happen?

   (1) The plate absorbed microwaves during heating.
   (2) The heated pizza warmed the plate.
   (3) The oven was not working properly.
   (4) The air in the oven was being heated.
   (5) The air in the plate was being heated.

3. Why is a microwave oven more energy efficient than a conventional oven?

   (1) A microwave oven heats only food.
   (2) A microwave oven does not use electricity.
   (3) A microwave oven neats the outside of food more quickly than the inside.
   (4) A microwave oven does not glow.
   (5) A microwave oven does not feel hot.

Question 4 refers to the following illustration and passage.

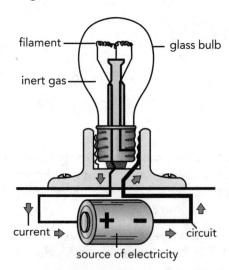

filament — glass bulb

inert gas —

current ➡ ⬅ circuit

source of electricity

The scientific name for a common light bulb is *incandescent lamp.* When this light is on, an electric current flows through a thin tungsten wire called a filament. Because tungsten is highly resistant to the flow of electrons, the filament is heated to about 5,400°F as the current passes through it. The hot filament glows, giving off both heat and light.

An incandescent lamp is filled with an inert gas that does not react with tungsten. The inert gas prevents the filament from burning out. Filaments do tend to become brittle and to break with long-term use. When a filament breaks, the light goes out—a condition most people refer to as "a bulb (or light) burning out."

4. The wires carrying the electric current from the battery to the bulb are made of copper. What can you infer to be true about copper?

    (1) Copper cannot be wound into a spiral shape in the way tungsten can.
    (2) Copper wire is less expensive than tungsten wire.
    (3) Copper is less resistant to the flow of electrons than is tungsten.
    (4) Copper is more resistant to the flow of electrons than is tungsten.
    (5) Copper turns green when exposed to moist air.

Questions 5 and 6 refer to the following diagrams.

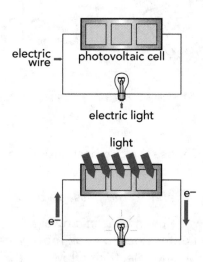

electric wire — photovoltaic cell

electric light

light

e⁻   e⁻

5. What is the purpose of a photovoltaic cell?

    (1) to change chemical energy into electrical energy
    (2) to change electrical energy into light energy
    (3) to change light energy into electrical energy
    (4) to change light energy into chemical energy
    (5) to change chemical energy into light energy

6. Which of the following devices uses a photovoltaic cell?

    (1) clinical thermometer
    (2) electric train
    (3) microwave oven
    (4) sunglasses
    (5) solar-powered calculator

Questions 7–9 refer to the following illustrations and passage.

Flashhlight
random frequencies, random directions

Laser
single frequency, focused beam

A laser is a special device that produces and emits a highly focused beam of light of a single frequency (color). Often called "pure light," laser light is *coherent light*, meaning its light waves move in step with one another so that the crest of one wave coincides with the crest of each other wave. The amount and the frequency of coherent light emitted by a laser depend on the design of the laser.

Coherent light can be thought of as a beam of single-energy photons (packets of light energy). Coherent waves can be made intense and can be highly focused in a single direction.

Light from a flashlight is incoherent light. *Incoherent light* consists of many frequencies (photons of different energy) traveling in a beam that cannot be intensified or highly focused in any single direction.

Lasers have many applications.

• In medicine, lasers are used to repair tears in the retina of the human eye. Laser light can repair damaged tissues without affecting surrounding tissue. Also, during surgery laser light is used to close cut blood vessels and to bore holes in the human skull.

• In communications, lasers are used to transmit data. Optical fibers transmit laser light signals in telephone and computer systems. Laser light is also used to play audio compact discs and videodiscs.

• In metallurgy, lasers are used to bore holes in metal and to weld different metals together.

7. Which of the following does <u>not</u> describe light from a flashlight?

(1) coherent
(2) multidirectional
(3) light waves of many different frequencies
(4) photons of many different energies
(5) incoherent

8. For which of the following applications would laser light be most useful?

(1) headlights of a car
(2) overhead lighting in a gymnasium
(3) light emitted from a computer monitor
(4) a signal sent from Earth to the Moon
(5) a signal emitted by a car security alarm

9. Which of the following can you infer from the passage?

(1) The energy output of a laser depends on the type of material that the laser beam strikes.
(2) The energy output of a laser depends primarily on the type of materials from which the laser is constructed.
(3) The energy output of a laser is unlimited.
(4) The energy output of a single laser can be varied over a wide range.
(5) The energy output of a laser does not depend on the type of laser being used.

Questions 10 and 11 refer to the following flowchart.

MAIN COMPONENTS OF A COMPUTER

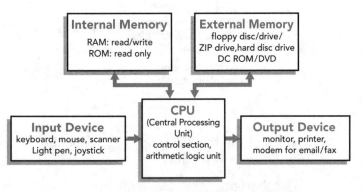

10. A computer system is often compared to a human body. In this comparison what subsystem of a computer is most likely compared to the human excretory system?

    (1) internal memory
    (2) external memory
    (3) input device
    (4) central processing unit
    (5) output device

11. When a computer is directed to multiply two numbers, where is this calculation performed?

    (1) internal memory
    (2) external memory
    (3) input device
    (4) central processing unit
    (5) output device

Question 12 refers to the following passage.

Nearly 150 years ago, the fastest horses in the country worked for the Pony Express. Their job was to carry mail from Missouri to California. Running from before sunrise to after sunset, mail could be delivered in an astounding ten days. Today, by means of a desktop computer, electronic mail (e-mail) can be sent in a few seconds.

At the heart of every computer are microchips, tiny electronic circuits that code and process all information in numerical form. While the first computers used vacuum tubes and filled a room, today's computers easily fit on a desk, on your lap, or even in your hand. Modern microchips are no larger than your fingernail.

Over the last few decades computers have gotten smaller, faster, and more powerful. There is more computing power (ability to do multiple, complex functions) in a single desktop computer today than was contained in all of the computers together that NASA used to send astronauts to the Moon and back in 1969!

12. Which of the following is <u>not</u> mentioned as an advantage of today's computers over earlier models?

    (1) smaller size
    (2) use of microchips
    (3) lower cost
    (4) faster operating speed
    (5) increased power

Answers are on page 108.

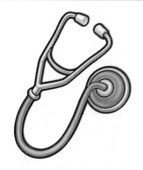

Science in Personal and Social Perspectives

GED Science pages 115–138

Directions: Choose the <u>one best answer</u> to each question.

Questions 1 and 2 refer to the following passage.

At the beginning of the twenty-first century, there is new hope for cancer patients. New drugs, such as Gleevec, fight cancer in a novel way. Taken as a pill, Gleevec targets specific leukemia cells for destruction and leaves healthy cells alone. In early testing, Gleevec is showing remarkable promise in stopping the progression of this type of cancer.

Before the invention of targeted cancer-cell fighters, standard treatments for cancer were surgery, radiation, and chemotherapy. Surgery almost always leaves behind cancer cells, and radiation and chemotherapy each can kill more healthy cells than cancer cells. What's more, these traditional cancer treatments are not effective against many types of cancer.

The hope now is to develop other cancer target-seeking drugs that will be able to seek out and destroy various types of cancer cells before the cancer harms the patient.

1. Which phrase does <u>not</u> describe Gleevec?

    (1) part of a new strategy in fighting cancer
    (2) can be taken in pill form
    (3) leaves healthy cells alone
    (4) able to identify cancer cells
    (5) a sure cure for many cancers

2. What may be one advantage of radiation?

    (1) known medicines available for nausea
    (2) treatment covered by insurance
    (3) knowledge of long-term effects
    (4) statistical probability for total cure
    (5) known destruction of healthy cells

Question 3 refers to the following passage.

Alzheimer's disease is a progressive brain disorder that starts with a gradual memory loss. As the disease progresses, the patient loses language, perceptual, and motor skills. Eventually Alzheimer's patients lose the ability to care for themselves.

Patients with Alzheimer's disease may live for many years after the onset of the symptoms. The average time from the onset to the time of death is five to ten years. Death usually results from disorders such as pneumonia, from which Alzheimer's patients have a difficult time of recovering.

No one knows what causes Alzheimer's disease. It is known that Alzheimer's is most common in older patients and that family history plays a role. Also, the disease involves distinctive formations in brain cells. These formations cause brain cells to shrink and die, leaving gaps in the brain's messaging network. The inability of the brain to function in a normal way gives rise to the symptoms of Alzheimer's.

3. Which of the following results of Alzheimer's disease <u>cannot</u> be inferred from the passage?

    (1) the uncertainty about the patient's future
    (2) the likely death of the patient within ten years
    (3) the cost of expensive medicines for the patient
    (4) the emotional difficulties that family members experience
    (5) the extent of personal care needed by the patient

Questions 4–7 refer to the following passage.

Medical researchers have discovered that smoking during pregnancy can be very harmful to a developing fetus. The fetus of a mother who smokes two packs of cigarettes each day loses 40 percent of the oxygen that would normally be present. This decrease in oxygen causes a baby born to a mother who smokes to weigh about six to seven ounces lighter than a baby born to a nonsmoker.

The risk of having a stillborn child increases for a mother who smokes. Smoking one pack each day increases fetal death rate by 20 percent. Smoking two or more packs each day increases fetal death rate by more than 35 percent. Mothers who smoke are twice as likely to have miscarriages and are more likely to suffer other complications than are nonsmokers during pregnancy, including premature births.

Smoking at any time during pregnancy also increases the chance that a fetus will develop a malformed heart or other organ. Also, smoking doubles the chance for *abruptio placentae*—a premature separation of the placenta from the uterine wall, which may result in the death of the fetus.

The potential for harm to the child of a smoking mother does not end at birth. Young children of smokers are about 50 percent more likely to die of crib death and twice as likely to develop a lung illness or respiratory allergies than are children of nonsmokers.

4. On average, how much lighter is the newborn of a woman who smokes during pregnancy than is the newborn of a nonsmoker?

   (1) less than two ounces
   (2) two to three ounces
   (3) four to five ounces
   (4) six to seven ounces
   (5) eight or more ounces

5. What is *abruptio placentae*?

   (1) another term for sudden infant death syndrome, or crib death
   (2) a complication occurring during the first three months of pregnancy
   (3) a birth defect in the fetus
   (4) a diminished supply of oxygen to the fetus due to smoking
   (5) premature separation of the placenta from the uterine wall

6. Which of the following is <u>not</u> mentioned as a possible adverse effect of smoking while pregnant?

   (1) miscarriage
   (2) premature birth
   (3) brain damage
   (4) low birth weight
   (5) heart malformation

7. Why do babies born to mothers who smoke have a lower than average birth weight?

   (1) the mother's poor eating habits
   (2) the presence of nicotine in the mother's blood
   (3) an insufficient supply of oxygen to the fetus
   (4) the increased risk of *abruptio placentae*
   (5) the reduced health of the mother due to smoking

Questions 8 and 9 refer to the table below.

The following table shows the risk factor of various birth-control methods when instructions are carefully followed. All methods are even less effective when not used properly.

BIRTH-CONTROL METHOD	RISK FACTOR *(approximate percent of women who become pregnant each year while using this method)*
male condom	16%
female condom	21%
female diaphragm or cervical cap	18%
spermicide	30%
IUD	4%
birth-control pill	6%
morning-after pill	25%
sterilization (tubal ligation or vasectomy)	almost 0%
rhythm method	19%
withdrawal method	24%
abstinence	0%

8. Of the following birth-control methods listed in the table, which is the most effective?

   (1) male condom
   (2) spermicide
   (3) birth-control pill
   (4) sterilization
   (5) withdrawal method

9. In which of the following publications or advertisements would you be <u>least</u> likely to find this table?

   (1) an advertisement for female condoms
   (2) a community health-clinic brochure
   (3) a book on different forms of birth control
   (4) a book on women's health
   (5) an advertisement for birth-control pills

Question 10 refers to the following passage.

Saccharin is a chemical substance that is about 500 times sweeter than ordinary sugar. Saccharin is used as a sugar substitute in many foods and beverages. In 1978 following the discovery that saccharin increases the incidence of bladder cancer in rats, the U.S. Food and Drug Administration (FDA) began requiring that all products containing saccharin carry a warning label.

10. Which comparative study could provide evidence that saccharin increases the risk of bladder cancer in humans?

    A. a study of the exercise habits of a group of people who are presently bladder-cancer patients
    B. a study of the long-term health of a group of presently healthy people who do not eat saccharin
    C. a study of the long-term health of a group of presently healthy people who eat saccharin regularly
    D. a study of the present eating habits of a group of people who are bladder-cancer patients

    (1) comparing the results of studies A and B
    (2) comparing the results of studies A and C
    (3) comparing the results of studies B and C
    (4) comparing the results of studies B and D
    (5) comparing the results of studies C and D

11. A dentist wants to prevent the possible spread of AIDS in her dental office. Which of the following steps is the <u>least</u> practical precaution she can take?

    (1) wear disposable plastic gloves
    (2) refuse to do dental work that causes gums to bleed
    (3) disinfect all dental instruments before each use
    (4) place all blood-soaked items in a plastic bag for safe disposal
    (5) wash her hands before seeing each new patient

Questions 12–14 refer to the following passage.

A controversial health issue today is the use of irradiation to preserve food. In this process high-energy gamma rays (obtained from the radioactive decay of cobalt) or high-energy X rays are used to kill mold, insects, such parasites as trichina worms in pork, and bacteria such as *Salmonella*.

Food irradiation has been used in dozens of countries, but it has only slowly become used in the United States. The U.S. Food and Drug Administration (FDA) first approved irradiation for harvested wheat and potatoes in the early 1960s. No other foods were approved until the 1980s, when spices, pork, fruits, and vegetables were added to the list. Most recently, beef was added to the approved list in 1997. Many other foods are expected to be approved for irradiation in the coming years.

Although gamma rays are a form of nuclear radiation, irradiated food is not radioactive. It can be eaten immediately, or it can be vacuum-packed and stored safely for several years. Supporters claim that irradiation is a safe alternative to pesticides and more traditional preservatives.

However, critics point out that because irradiation can be used only on harvested crops, pesticides will still be needed to protect crops in the fields. What's more, they say that irradiation may kill odor-producing organisms that signal spoilage without killing the sources of food poisoning. Thus, irradiation may destroy an important natural warning system.

Some critics also have claimed that irradiation may change the chemical makeup of food and create carcinogens (cancer-causing chemicals). Supporters, on the other hand, contend that an equal amount of carcinogens is produced when food is cooked or frozen.

Both supporters and critics agree that, with regard to the handling of food—irradiated or not—both strict sanitation and cooking standards are essential if public health is to be safeguarded.

12. Which is the best summary of the passage?

(1) Food irradiation is an accepted method of food preservation the world over.
(2) Although controversial, the use of food irradiation is increasing in the United States.
(3) Food irradiation does not result in deadly radioactive waste.
(4) Food irradiation can bring down the costs of dried foods.
(5) Because it produces high levels of carcinogens, food irradiation should not be allowed by the FDA.

13. When fruit is irradiated for a very long time, it turns squishy. What is one possible effect on fruit of long exposure to gamma rays?

(1) the breaking down of the fruit's chromosomes
(2) an altering of the flavor of the fruit's juice
(3) the breaking down of the fruit's cell walls
(4) the changing the fruit's color
(5) the destruction of the fruit's cellular nuclei

14. Which of the following facts is <u>least</u> important to FDA food scientists who are trying to determine what, if any, are the long-term dangers of food irradiation?

(1) The complete chemical makeup of any food is impossible to measure.
(2) So far, no adverse reactions to irradiation have been observed.
(3) Cancer associated with food irradiation may take twenty or more years to develop.
(4) The exact chemical changes caused by gamma radiation are impossible to determine.
(5) Because of the well-publicized dangers of radioactive waste, many people fear the food irradiation process.

Answers are on page 108.

History and Nature of Science

GED Science pages 139–155

Directions: Choose the <u>one best answer</u> to each question.

Questions 1 and 2 refer to the following passage.

Archimedes (287–212 B.C.) was a Greek mathematician and inventor who is best known today for his discovery of the law of hydrostatics. This law states that a body immersed in a liquid loses weight equal to the weight of the liquid it displaces (takes the place of). Archimedes very likely discovered this law while taking a bath!

1. Which fact would Archimedes <u>not</u> have used to help explain the law of hydrostatics?

   (1) While walking neck-deep in water, a person feels almost weightless.
   (2) Light objects, such as wood, float while heavy objects, such as iron, sink.
   (3) A heavier floating object displaces more water than a lighter floating object.
   (4) A rock on a river's bottom is easier to lift than the same rock on land.
   (5) The weight of a closed container of water does not change as the container is slowly heated.

2. In which of the following is Archimedes's discovery an important factor?

   A. the design of a submarine
   B. the anatomy of a fish
   C. the design of a space rocket

   (1) A only
   (2) B only
   (3) C only
   (4) Both A and B
   (5) Both A and C

Question 3 refers to the following illustration.

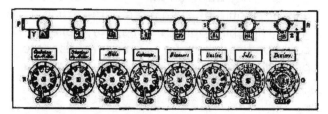

3. In 1642 Blaise Pascal developed the machine shown above for his father, a tax collector. What is the most likely name and purpose of this machine?

   (1) a mechanical clock, used to tell time
   (2) a game machine, used to play chess
   (3) a calculating machine, used to do math
   (4) a kitchen tool, used to sharpen knives
   (5) a shop tool, used to cut gears

Question 4 refers to the illustration below.

CATAPULT

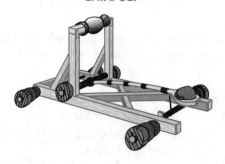

4. Invented more than 2,000 years ago, the catapult was in use well into the Middle Ages (about 500 years ago). For what activity was the catapult designed?

   (1) agriculture
   (2) warfare
   (3) entertainment
   (4) overland travel
   (5) water travel

Questions 5 and 6 refer to the passage below.

Alchemy, a quest to change common metals into gold, flourished during the Middle Ages. Alchemists believed that nature itself formed gold out of less perfect metals. The hope of alchemists was to learn to copy the gold-making process that occurred naturally on Earth.

As part of their quest, alchemists searched for a substance called the "philosopher's stone." They believed that such a stone, if it existed, would be even more perfect than gold. The philosopher's stone, they believed, could be used to make common metals more perfect, more gold-like. How this would occur is not known!

5. What belief, proposed 1,000 years before the Middle Ages, did alchemists most likely think was true?

   (1) Archimedes's belief that immersed objects feel lighter in weight.
   (2) Aristotle's belief that all things tend to reach perfection.
   (3) Ptolemy's belief that Earth is at the center of the solar system.
   (4) Pythagoras's belief that all things are made of four elements.
   (5) Erasistratus's belief that nerves carry "nervous spirit" from the brain.

6. Which of the following facts most likely led alchemists to believe in the perfection of gold?

   (1) Gold can be shined and does not tarnish as do other metals.
   (2) Gold is scarce and does not occur as large pieces or chunks.
   (3) Gold is a very soft metal, even softer than copper or aluminum.
   (4) Gold sinks when placed in water.
   (5) Gold can be melted and mixed with other metals such as copper and zinc.

Question 7 refers to the following illustration.

7. In about 1752 Benjamin Franklin performed the experiment illustrated above. What was Franklin trying to learn more about?

   (1) the strength of kite string
   (2) the flying properties of kites
   (3) the prediction of weather
   (4) the nature of electricity
   (5) the strength of gravity

Question 8 refers to the illustration below.

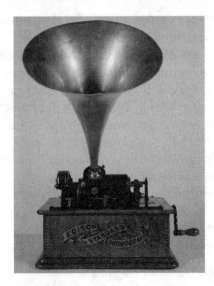

8. Thomas Edison invented this device in 1877. What is its name?

   (1) Edison's telephone
   (2) Edison's electric light
   (3) Edison's television
   (4) Edison's phonograph
   (5) Edison's computer

Questions 9 and 10 refer to the cartoon below.

WAY TOO GENETICALLY ENGINEERED CHICKEN

9. What type of modern research does the cartoon most likely call into question?

   (1) repairing defective genes in a fetus
   (2) placing genes of one species into other species
   (3) determining the genetic structure of living organisms
   (4) attempting to understand inherited characteristics by studying genes
   (5) using genetic testing as a way of identifying parent organisms

10. With which statement would the cartoonist most likely agree?

   (1) Scientists should be financially supported in all research efforts.
   (2) The creation of new species is an exciting new scientific possibility.
   (3) There is more to good science than trying to do everything that is possible.
   (4) Scientific research should be allowed to proceed in all possible directions.
   (5) Exciting research is now being done in all areas of genetic engineering.

Questions 11 and 12 refer to the following passage.

Pythagoras, a sixth-century B.C. Greek mathematician, believed that the secrets of the universe were revealed by numbers. Pythagoras based this conclusion on a discovery about musical harmony.

Suppose you take two strings whose lengths are in the ratio of two small whole numbers, say 2:3. You can do this by choosing one string 2 feet long and another string 3 feet long. If you tightly stretch and then pluck the strings, you will find that the two sounds are harmonious—pleasant to hear at the same time.

Pythagoras concluded that whole number ratios, such as 2:3, have much to do with the way the world is structured. This conclusion led Pythagoras to believe that heavenly bodies (all distant objects that we observe in the sky) are separated from one another by distances that are in harmonic ratios—ratios of lengths that produce pleasing sounds in stretched strings. He also believed that the movement of heavenly bodies made a harmony that he called the "music of the spheres."

11. How would Pythagoras most likely describe the music of the spheres?

   (1) a violin-like sound
   (2) a loud sound
   (3) a delightful sound
   (4) an imaginary sound
   (5) a sound like plucked strings

12. What insight was shown by Pythagoras, even though his conclusion about the music of the spheres was not correct?

   (1) Heavenly bodies move.
   (2) Mathematics is important in science.
   (3) Moving heavenly bodies make sounds.
   (4) Vibrating strings make sounds.
   (5) Musical harmony is pleasant to hear.

Questions 13–16 refer to the illustration and passage below.

PARTICLE-LIKE NATURE OF LIGHT

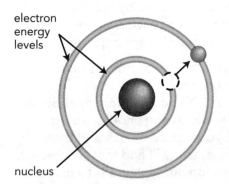

electron energy levels

A photon is absorbed by an electron, resulting in the electron moving to a higher energy level in the atom.

nucleus

WAVE-LIKE NATURE OF LIGHT

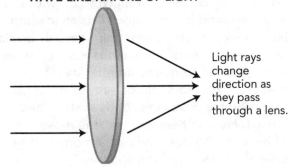

Light rays change direction as they pass through a lens.

Light, the form of energy that has made life possible on Earth, provides scientists with one of their greatest mysteries. Just what is light? Is it a particle (a packet of energy of a certain size) or is it a wave (a vibrating passage of energy)?

Scientists have discovered that sometimes light has particle-like features, and sometimes it has wave-like features. It depends on what the light is doing.

- When emitted or absorbed by an atom, light acts similar to a particle and is called a *photon*. Unlike other particles, a photon has no mass and is always moving at the speed of light—until it is absorbed. The energy of a photon can be measured, but the size of a photon cannot.

- When moving through a lens, light refracts (changes direction) similar to a wave and is called an electromagnetic wave.

In the next century scientists may have a better understanding of electromagnetic energy (light). But for now, the present understanding is called *wave-particle duality*—acknowledging both the wave and particle features of light.

13. What name is given to a packet of electromagnetic energy?

    (1) ion
    (2) electron
    (3) wave
    (4) photon
    (5) particle

14. What is the most unusual particle-like feature of light?

    (1) speed of movement
    (2) interaction with matter
    (3) lack of mass
    (4) energy-carrying property
    (5) ability to travel through empty space

15. Light changes direction as it moves from air to water. What property of light is demonstrated by this fact?

    (1) the wave nature of light
    (2) the particle nature of light
    (3) both the wave and particle nature of light
    (4) the energy of light
    (5) the speed of light

16. What property of a scientific theory is best pointed out by this passage?

    (1) A scientific theory can never account for all available evidence.
    (2) A scientific theory does not need to account for all available evidence.
    (3) A scientific theory may consist of more than one model to account for all available evidence.
    (4) A scientific theory must account for all available evidence by way of a single model.
    (5) A scientific theory does not need to be supported by evidence of any kind.

Answers are on page 109.

Plant and Animal Science

GED Science pages 159–220
Complete GED pages 459–496

Directions: Choose the <u>one best answer</u> to each question.

Questions 1–3 refer to the graph at the right.

The graph shows the amount of light absorbed by chlorophyll at various wavelengths. Chlorophyll is the pigment in plant leaves that absorbs sunlight for use in photosynthesis. It also gives leaves their green color.

LIGHT ABSORPTION OF CHLOROPHYLL

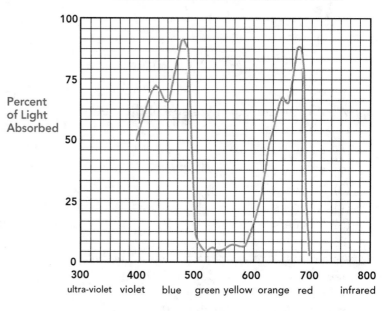

Wavelength of Light (in nanometers)

1. Which two colors of light are absorbed most by chlorophyll?

   (1) blue and green
   (2) green and yellow
   (3) violet and yellow
   (4) blue and orange
   (5) green and red

2. Which stationary organism would be <u>least</u> likely to have a light-absorption graph similar to the graph above?

   (1) broccoli
   (2) mushrooms
   (3) lettuce
   (4) lawn grass
   (5) ferns

3. Color is determined by light that is reflected (not absorbed). From the graph you can conclude that the color of chlorophyll results mainly from reflected light of which wavelengths?

   A. 400 to 500 nanometers
   B. 500 to 600 nanometers
   C. 600 to 700 nanometers

   (1) A only
   (2) B only
   (3) C only
   (4) both A and B
   (5) both A and C

Question 4 refers to the following passage and diagram.

Each year of its life, a tree grows by adding a new layer of cells to the outside of its stem (trunk). As shown below, these layers form a pattern of growth rings that can be seen by looking at a cross section of the tree.

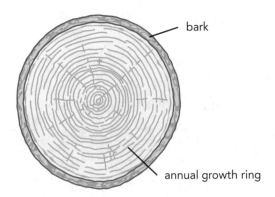

bark

annual growth ring

4. What is the approximate age of the tree shown above?

   (1) 5 years
   (2) 13 years
   (3) 20 years
   (4) 34 years
   (5) 47 years

5. As a general rule, animals living in the wild have a shorter life expectancy than animals raised in captivity. What is the most likely reason for this difference?

   (1) Animals in the wild are more likely to be killed by predators.
   (2) Animals in the wild eat nonnutritional foods.
   (3) Animals in the wild age more quickly.
   (4) Animals in the wild have too much room in which to exercise.
   (5) Animals in the wild have more stress.

6. The fact that cool water contains more dissolved oxygen than warm water is important to many water animals. Trout, for example, survive best in water that is shaded and kept cool by overhanging branches. Trout can die in water that is constantly warmed by sunshine.

   What is the best explanation for the fact that trout do not thrive in warm water?

   (1) Warm water contains less nutrients for trout than cool water.
   (2) Warm water contains less oxygen for trout than cool water.
   (3) Warm water contains more oxygen for trout than cool water.
   (4) Warm water flows more quickly than cool water, which makes it difficult for trout to swim.
   (5) Warm water carries odors better, which enables predators to easily find trout.

Question 7 refers to the diagram below.

FERTILE CHICKEN EGG

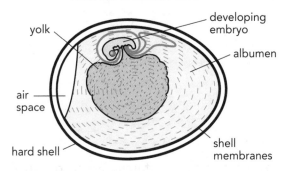

yolk

developing embryo

albumen

air space

hard shell

shell membranes

7. Which of the following can you infer from the diagram above?

   (1) As a chicken embryo gets larger, the yolk gets smaller.
   (2) As a chicken embryo gets larger, the albumen gets larger.
   (3) Albumen makes up most of the weight of a chicken egg.
   (4) The oxygen supply for a developing embryo comes from an air space within its shell.
   (5) Except for size, a chicken egg is identical to a duck egg.

Questions 8–11 refer to the following information and graph.

A wildlife biologist placed a number of young trout in a lake that previously had no trout. The graph below shows the growth of this trout population over many years. Also shown is the estimated trout population that may have grown in the absence of population-limiting factors—things that kill trout before they reach the end of their natural life span.

TROUT POPULATION IN A LAKE

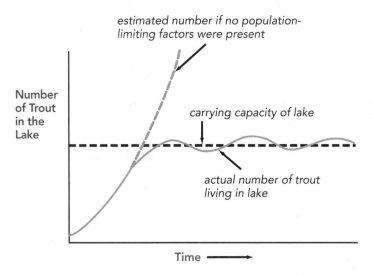

8. Which of the following would <u>least</u> likely be classified as a population-limiting factor?

   (1) water pollution in the lake
   (2) bacteria in the lake that cause fish diseases
   (3) a lack of adequate food resources in the lake for trout
   (4) the average natural life span of trout that live in the lake
   (5) the presence of larger fish in the lake that feed on trout

9. Which of the following can you infer to be true about population-limiting factors?

   (1) They tend to be greater during winter months than summer months.
   (2) They lead to a relatively stable number of trout in the lake.
   (3) They depend on the size and location of the lake.
   (4) They decrease as the number of trout increases.
   (5) They affect trout more than they affect smaller fish.

10. What does *carrying capacity* mean?

   (1) the size of the trout population in the absence of population-limiting factors
   (2) the exact number of trout in the lake at any one time
   (3) the number of trout that the lake can support
   (4) the number of trout that die each year
   (5) the number of trout that are born each year

11. What would be the likely result if summer homes were built along the lakeshore?

   (1) The carrying capacity of the lake would likely decrease.
   (2) The number of population-limiting factors would likely decrease.
   (3) The size of the lake would likely decrease.
   (4) The carrying capacity of the lake would likely increase.
   (5) The number of trout living in the lake would likely increase.

Questions 12 and 13 refer to the following information.

Bacteria are often classified by whether they need oxygen to live.

- *Aerobic bacteria* can survive only in places where oxygen is plentiful. Simple bread mold is an example.

- *Anaerobic bacteria* are found only where oxygen is limited or entirely absent. An example is tetanus, which can live within puncture wounds.

- *Facultative anaerobes* can survive with or without oxygen. *Escherichia coli*, which is present in the human digestive tract, is an example.

12. In which of the following places would aerobic bacteria be able to survive?

    A. in fermenting wine
    B. in infected human lungs
    C. on a piece of cheese

    (1) A only
    (2) B only
    (3) C only
    (4) Both A and C
    (5) Both B and C

13. What type(s) of bacteria might be found in the digestive tracts of animals?

    (1) aerobic
    (2) aerobic and anaerobic
    (3) aerobic and facultative
    (4) anaerobic and facultative
    (5) aerobic, anaerobic, and facultative

Question 14 refers to the passage and illustration below.

Most bacteria divide by a process called *binary fission*. In binary fission a bacterium cell first produces a copy of its own DNA. Then the cell grows and splits into two parts, each part receiving a complete copy of the DNA.

BINARY FISSION

❶ The cell grows and prepairs to divide.

❸ The DNA and its copy separate as the cell grows and divides.

❷ The DNA makes a copy of itself.

❹ The cell splits into two identical cells. Each cell has a copy of the original DNA.

14. Which of the following does <u>not</u> occur during binary fission?

    (1) cell growth
    (2) the copying of DNA
    (3) cell splitting
    (4) genetic diversity
    (5) an increased number of cells

15. On the average, 1 cubic foot of air contains about 100 bacteria. About how many airborne bacteria are in a classroom that is 60 feet long, 50 feet wide, and 10 feet high?

    (volume = length × width × height)

    (1) 30,000
    (2) 300,000
    (3) 3,000,000
    (4) 30,000,000
    (5) 300,000,000

Questions 16 and 17 refer to the information below.

A plant scientist created a new hybrid grass by crossing a desert grass with a shade grass. To determine the hybrid's ideal growing conditions, an experiment will be performed in which the hybrid is grown in four planters. Each planter will be given the amounts of sunshine and water shown in the diagram below. The average height of the grass in each planter will be recorded weekly.

PLANTERS CONTAINING HYBRID GRASS SEEDS

A

Watered three times each week

Receives 8 hours or more sunshine each day

B

Watered three times each week

Receives 4 hours or less sunshine each day

C

Watered once every two weeks

Receives 8 hours or more sunshine each day

D

Watered once every two weeks

Receives 4 hours or less sunshine each day

16. Which of the following conditions, if not met, would <u>least</u> affect the results of the experiment?

 (1) The depth of the soil should be identical in all four planters.
 (2) The drainage holes should be identical in all four planters.
 (3) All four planters should be kept on the same table.
 (4) The same type of soil should be placed in all four planters.
 (5) Each planter should be kept free of weeds and insects.

17. Which two planters should be compared in order to determine the effect of water on the growth of the hybrid grass when it receives full days of sunshine?

 (1) A and B
 (2) A and C
 (3) B and C
 (4) B and D
 (5) C and D

Questions 18–21 refer to the following graph.

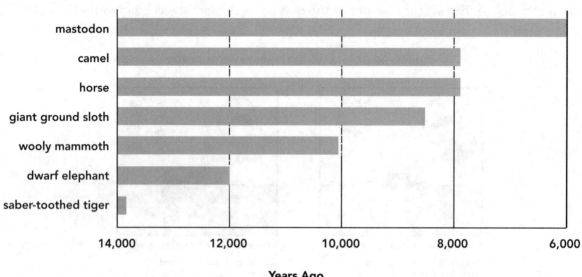

RECENT EXTINCTION OF LARGE MAMMALS ONCE NATIVE TO NORTH AMERICA

Years Ago

18. Which of the following is a restatement of information from the graph above?

    (1) For most of its existence, the mastodon competed with the woolly mammoth for food.
    (2) Of the animals listed the saber-toothed tiger was the last to become extinct in North America.
    (3) The camel and horse became extinct in North America at about the same time.
    (4) The giant ground sloth had a longer life expectancy than the dwarf elephant.
    (5) Mammals are more likely to become extinct than are reptiles.

19. About how many years passed between the extinction of the wooly mammoth and the extinction of the mastodon?

    (1) 4,000
    (2) 6,000
    (3) 8,000
    (4) 10,000
    (5) 12,000

20. Which of the following factors is <u>least</u> likely to lead to the extinction of an animal species in North America?

    (1) the disappearance of food resources
    (2) an increasing number of predators
    (3) an extreme temperature change
    (4) an increase in air-pollution levels
    (5) the extinction of similar animals in Europe

21. Which of the following animals is once again commonly found throughout North America?

    (1) dwarf elephant
    (2) horse
    (3) saber-toothed tiger
    (4) camel
    (5) mastodon

Questions 22–26 refer to the following information.

In a learning-theory experiment a group of octopuses was trained to attack a red ball. Another group was trained to attack a white ball. Meanwhile, two groups of untrained octopuses watched: One group watched the red-ball tank; the other watched the white-ball tank. When both red and white balls were placed in the tanks with the untrained octopuses, each octopus attacked only the color ball that was attacked in the group it had watched. Although octopuses have the largest brain of any invertebrate, this result surprised many scientists. This social-learning skill had been considered to be beyond the ability of an octopus.

The statements in questions 22–26 can be classified into one of these five categories:

- *experiment*—a procedure used to investigate a problem

- *finding*—an experimental result or a conclusion reached as part of the investigation

- *hypothesis*—a reasonable, but not proved, explanation of an observed fact

- *prediction*—an opinion about something that may occur in the future

- *nonessential fact*—a fact that does not directly help the researcher understand the problem being investigated

22. Octopuses that observed a red ball being attacked chose to attack a red ball themselves.

    How is this statement best classified?

    (1) experiment
    (2) finding
    (3) hypothesis
    (4) prediction
    (5) nonessential fact

23. An octopus has a soft, oval body and eight arms and lives mostly at the bottom of the ocean.

    How is this statement best classified?

    (1) experiment
    (2) finding
    (3) hypothesis
    (4) prediction
    (5) nonessential fact

24. With further research, scientists will find that octopuses have a wide range of learning abilities not yet discovered.

    How is this statement best classified?

    (1) experiment
    (2) finding
    (3) hypothesis
    (4) prediction
    (5) nonessential fact

25. To see if octopuses can learn by observation, untrained octopuses were allowed to watch trained octopuses attack colored balls.

    How is this statement best classified?

    (1) experiment
    (2) finding
    (3) hypothesis
    (4) prediction
    (5) nonessential fact

26. The learning ability of an octopus most likely arose from its need to learn from other octopuses at an early age, because the parents of an octopus die when it is hatched.

    How is this statement best classified?

    (1) experiment
    (2) finding
    (3) hypothesis
    (4) prediction
    (5) nonessential fact

Questions 27–29 refer to the following passage.

Important characteristics of the six most familiar groups into which most animals can be classified are given below.

Fish have two organs that no other animals have: gills and an air bladder. Fish use gills to take dissolved oxygen out of water. The air bladder, a thin-walled sac that acts as a float, controls the level at which fish swim. To reproduce, fish deposit soft eggs on the bottom of the body of water in which they live.

Insects are many-legged creatures whose bodies are divided into three distinct regions. Many insects have wings used for flying. Other insects have neither legs nor wings. Most insects go through a four-stage development process in which the young look nothing like the adult. To reproduce, insects lay eggs.

Amphibians live for a time in water and for a time on land. Young amphibians live only in water. As they mature, though, they develop features that allow them to live on land. However, amphibians do not become total land animals. Like fish, amphibians must reproduce and lay eggs in water.

Reptiles are land animals. Unlike fish and amphibians, they lay an amniotic egg—an egg that is surrounded by a protective membrane and tough shell. Reptiles are also characterized by a dry body covering of some type of horny scales or plates.

Birds are the only type of animal that are well adapted to life in the air, in the water, and on land. Flight is accomplished by the movement of a pair of feather-covered wings. Birds, like reptiles, lay amniotic eggs.

Mammals have the most highly developed organ systems and brain. Most mammals give birth to live young, offspring that look similar to their parents from the moment of birth. Young mammals are nourished with milk from the mammary glands of their mother.

27. A salamander is a timid, harmless animal that has a thin, elongated body and a long tail. Most species have four short legs. Salamanders live near streams and ponds in which the female lays her soft eggs. The larvae (the young form of a salamander) grow in the water, and they eventually develop lungs that allow them to live on land.

    How are salamanders classified?

    (1) fish
    (2) amphibians
    (3) reptiles
    (4) birds
    (5) mammals

28. The killdeer gets its name from the sound it makes—it seems to be saying "kill-dee." Killdeers build their nests in fields, where the female usually lays four spotted eggs. They are often seen flying around fields where they feast on crop-destroying insects.

    How are killdeers classified?

    (1) fish
    (2) amphibians
    (3) reptiles
    (4) birds
    (5) mammals

29. The sea horse gets its name from the shape of its head, which looks similar to the head of a tiny horse. This five-inch-long animal lives only in water. It uses its tail to cling to plants or floating vegetation. The female lays about 200 eggs, which she deposits in a pouch on the underside of the male sea horse's body.

    How is a sea horse classified?

    (1) fish
    (2) amphibian
    (3) reptile
    (4) bird
    (5) mammal

Questions 30–34 refer to the following passage.

The orangutan is one the most intelligent of all land animals. Its name comes from the Malay words *orang* and *hutan,* which mean "man of the forest." The two Indonesian islands of Borneo and Sumatra are the only places where these "forest men" live.

The adult male orangutans of Borneo roam freely and do not stay close to any family unit. A few miles away on Sumatra, however, an adult male does stay close to his mate and offspring. Differences in male orangutan behavior are most likely the result of learned or possibly inherited behavior patterns that best ensure the survival of orangutans in each location.

On the island of Borneo, male orangutans are not needed to help ensure the health and safety of the females or the offspring. Males wander freely through the feeding ranges occupied by females and their young. The older, dominant males mate with any fertile, receptive female. On Borneo if a large male were to stay near his mate and offspring, he likely would have to compete for food and actually decrease his offspring's chance of survival.

Sumatra, however, contains leopards that prey on female and infant orangutans, and siamangs, tree-dwelling apes that compete with orangutans for food. For both of these reasons, the survival chance for female and infant orangutans increases if the males stay close.

30. On Sumatra which animal is known to hunt orangutans?

    (1) leopards
    (2) siamangs
    (3) other orangutans
    (4) snakes
    (5) lions

31. What threat to the health of orangutans does the passage imply may exist on Borneo?

    (1) leopards
    (2) siamangs
    (3) starvation
    (4) infertility
    (5) overpopulation

32. What general biological principle is exemplified by male orangutans in Borneo and Sumatra when showing different parenting behaviors?

    (1) camouflage
    (2) speciation
    (3) reproduction
    (4) reaction
    (5) adaptation

33. What do the parenting behaviors of the orangutans on Borneo and Sumatra have in common?

    (1) The purpose of each is to maintain closely bonded family units.
    (2) The purpose of each is to produce the greatest number of surviving offspring.
    (3) The purpose of each is to establish a social group of orangutans.
    (4) The purpose of each is to enable orangutans to successfully compete for food with siamangs.
    (5) The purpose of each is to protect orangutans from human hunters.

34. Which scientist's work would best help in the understanding of orangutan behavior?

    (1) Archimedes
    (2) Aristotle
    (3) Thomas Edison
    (4) Charles Darwin
    (5) Albert Einstein

Questions 35 and 36 refer to the following passage and diagram.

As shown below, the life cycle of a fern is divided into two parts: a sexual generation called the *gametophyte*, and an asexual generation called the *sporophyte*.

FERN LIFE CYCLE

After landing on moist soil, spores grow into new gametophytes.

Gametophyte

Female structure with egg cells

Male structure with sperm cells

Airborne spores are released by the sporophyte.

Sporophyte

A sperm cell unites with an egg cell.

A sporophyte develops from the fertilized egg.

35. What is the name given to the fern generation that contains both male structures and female structures?

(1) egg cells
(2) gametophyte
(3) sperm cells
(4) sporophyte
(5) spores

36. What is the best definition of spore?

(1) female egg cell
(2) male sperm cell
(3) fertilized egg cell
(4) sexual reproductive cell
(5) asexual reproductive cell

37. *Renewable resources* are substances in the environment that can be replaced as they are used. Food crops such as wheat and corn are renewable resources. *Nonrenewable resources* are substances such as petroleum (oil) and aluminum that cannot be replaced once they are gone.

Which of the following would most likely be classified as a renewable resource?

(1) coal
(2) iron ore
(3) diamonds
(4) forests
(5) natural gas

38. Seen from the top, trout are dark in color with a speckled appearance, similar to the bottom of a river or lake. Seen from the bottom, trout are light in color, similar to the water surface as seen by an underwater swimmer.

What name is given to the general scientific characteristic exemplified by the coloration of a trout?

(1) camouflage
(2) behavior
(3) hybridization
(4) convection
(5) stimulus-response

Questions 39–41 refer to the following drawing and passage.

ROUND DANCE

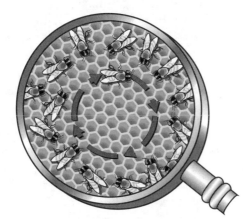

WAGGLE DANCE

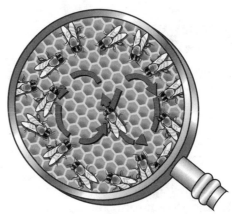

Among of the most interesting examples of animal communication are the round dance and waggle dance of honeybees.

After discovering flowers containing nectar, a honeybee returns to the hive and shares the news by performing one of two dances.

• A *round dance* is performed for nectar located within about 100 yards of the hive. The dancing bee moves in a circle, with other bees soon joining in as they smell the nectar on the dancing bee. The round dance does not tell either the direction or location of the nectar, only that it is close.

• A *waggle dance* is performed for nectar that is located farther from the hive than about 100 yards. The dancing bee moves in a figure eight, waggling (swaying back and forth) its abdomen along the straight part of the dance. The waggle dance tells both the direction and approximate distance of the nectar. The path the bee follows on the straight part indicates the direction of the nectar relative to the Sun. The number of figure-eight patterns that are repeated indicates the approximate distance of the nectar from the hive.

After seeing the round dance, other honeybees fly away from the hive in ever-widening circles until they find the nectar.

After seeing the waggle dance, other honeybees fly from the hive toward the nectar. Knowing the direction and approximate distance of the nectar, they can fly right to it.

39. For what purpose does a honeybee do the movement called the *waggle dance*?

(1) to share the smell of nectar
(2) to awaken other honeybees
(3) to indicate direction
(4) to indicate distance
(5) to indicate both distance and direction

40. What can you infer to be an advantage of the round dance over the waggle dance?

(1) The round dance involves participation of more than one bee.
(2) The round dance more accurately tells the distance of the nectar from the hive.
(3) The round dance takes less time.
(4) The round dance can be performed even when nectar is not close by.
(5) The round dance does not depend on the Sun being visible to the bees.

41. At what time on a clear, sunny day would the waggle dance be least effective?

(1) 8:00 A.M.
(2) 10:00 A.M.
(3) 1:30 P.M.
(4) 4:30 P.M.
(5) 9:30 P.M.

Question 42 refers to the following illustration.

HOW SKIN HEALS

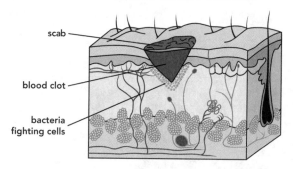

42. Which of the following is the <u>least</u> likely reason that a blood clot forms at a small cut in the skin?

(1) to help prevent further tearing of the skin
(2) to protect nearby hair follicles from an infected wound
(3) to prevent excessive blood loss from the wound
(4) to prevent bacteria from entering the wound
(5) to protect underlying tissue from further damage

Question 43 refers to the following graph.

PARAMECIUM GROWTH

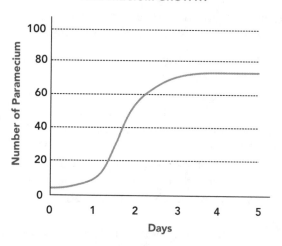

43. During which day was the rate of paramecium growth the greatest?

(1) day 1
(2) day 2
(3) day 3
(4) day 4
(5) day 5

Questions 44 and 45 refer to the following passage.

The Gaia hypothesis, proposed in the 1970s by British scientist James Lovelock, proposes that Earth behaves as a single living organism. According to this theory, Earth regulates its own temperature, provides itself with resources needed for life, disposes of its own wastes, and fights off disease.

The Gaia hypothesis is an extension of the concept of *homeostasis*. Homeostasis is a term referring to the tendency of a biological system to maintain a state of equilibrium, or balance. Individual organisms exhibit homeostasis; so do communities of organisms that form an ecosystem. The Gaia hypothesis takes the concept of homeostasis to the highest level in viewing Earth as a single self-regulating system.

44. Which of the following is <u>least</u> likely to be classified as an example of homeostasis?

(1) hormone regulation in animals
(2) cell growth in plants
(3) surrogate parenting
(4) predator-prey relationships
(5) warming effect of Earth's atmosphere

45. According to the Gaia hypothesis, which of the following would most likely harm Earth to the greatest degree?

(1) lightning-caused fires in the tropical rain forests
(2) severe earthquakes along major fault lines on each continent
(3) an ice age resulting from natural temperature variations
(4) collision with a large asteroid
(5) volcanic eruptions along the Ring of Fire

Answers are on page 109.

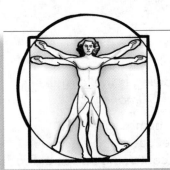

Human Biology

GED Science pages 221–253
Complete GED pages 459–496

Directions: Choose the <u>one best answer</u> to each question.

Questions 1 and 2 refer to the following diagram.

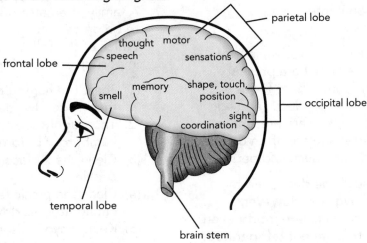

1. Which statement best summarizes the information in the diagram?

   (1) All human brains are about the same size.
   (2) Specific parts of a human brain control specific body functions.
   (3) Human intelligence is determined by brain size.
   (4) The surface of a human brain contains many folds.
   (5) On a human being the brain stem connects the spinal cord to the brain.

2. What part of the human brain is involved in vision?

   (1) frontal lobe
   (2) parietal lobe
   (3) temporal lobe
   (4) occipital lobe
   (5) brain stem

Question 3 refers to the following information.

Pain is unpleasant but does serve a useful purpose. Pain can alert us to an infection, and it can alert us to a harmful activity.

- A toothache may be a symptom of an infected tooth, which may get worse if not fixed.

- Back pain may be a sign that a person has been sitting or standing in a position that is stressful for the back. Without a change in posture, the pain may increase.

- A painful thumb after hitting it accidentally with a hammer is a sign that you may want to be more careful in the future!

3. Which type of pain listed below is a person <u>least</u> likely to complain about?

   (1) a sore neck from sleeping in a chair
   (2) sore muscles from exercising
   (3) a stomachache after eating too much
   (4) pain following major surgery
   (5) pain from touching a hot stove burner

Questions 4–8 refer to the passage below.

During the spring of 1984, an epidemic of food poisoning in the Midwest focused national attention on a disease known as salmonellosis. More than 18,000 people got sick from improperly pasteurized milk. At least one person died. Since then, there have been several smaller, but highly publicized, outbreaks of this same disease. These outbreaks involved hamburger meat, cantaloupes, and eggs.

Salmonellosis is caused by rod-shaped bacteria called *Salmonella* that are present in moist, high-protein foods. These foods include meat, poultry, milk products, and egg products. *Salmonella* bacteria are present in saliva and in fecal matter and are often carried by flies, other insects, and household pets.

Salmonella can usually be destroyed by cooking and by good hygiene. However, *Salmonella* can easily contaminate foods, even saved cooked foods, that are not refrigerated.

Food poisoning from *Salmonella* occurs most often during summer months and is often contracted at barbecues where food handling and hygiene may be careless. To reduce chances of *Salmonella* poisoning, refrigerate all perishable foods and make sure that people who handle food have clean hands.

4. Which of the following foods is <u>least</u> likely to carry the *Salmonella* bacteria?

    (1) oysters
    (2) duck eggs
    (3) cottage cheese
    (4) potato salad
    (5) salted crackers

5. During which month are most cases of salmonellosis in the United States most likely to occur?

    (1) May
    (2) July
    (3) November
    (4) January
    (5) March

6. Which of the following is the <u>least</u> likely possible source of *Salmonella*?

    (1) a plate of raw hamburgers at a summer barbecue
    (2) a picnic table containing scraps of food
    (3) flies buzzing around garbage cans
    (4) a campfire site
    (5) dirty baby diapers

7. A family at an afternoon barbecue wants to be protected against food poisoning. Which of the following is the <u>least</u> important concern for the family?

    (1) Toast the hamburger buns before making sandwiches.
    (2) Thoroughly cook chicken and hamburger before serving.
    (3) Clean the surface of the grill before cooking.
    (4) Clean the picnic table before setting eating utensils down.
    (5) Keep mayonnaise in a cooled container.

8. To prevent the possibility of food poisoning, which of the following should a shopper always check before buying a food item?

    (1) sugar content per serving
    (2) types of flavor additives it contains
    (3) date after which it should not be sold
    (4) number of calories per serving
    (5) date on which it was produced

Questions 9 and 10 are based on the following chart.

RECOMMENDED DAILY VITAMIN NEEDS OF MALE AND FEMALE ADULTS*

(ages 19–30)

	Males	Females
Vitamin A	1,000 μg	800 μg
Vitamin E	10 mg	8 mg
Vitamin K	75 μg	62 μg
Vitamin C	60 mg	60 mg
Thiamin	1.2 mg	1.1 mg
Riboflavin	1.3 mg	1.1 mg
Niacin	16 mg	14 mg
Vitamin B_6	1.3 mg	1.3 mg
Folate	400 μg	400 μg
Vitamin B_{12}	2.4 μg	2.4 μg
Vitamin D	5 μg	5 μg

mg = milligram; μg = microgram
* Based on figures from the National Academy of Sciences

9. Which of the following can be inferred from the chart above?

   (1) Vitamin A is more necessary than the other vitamins listed.
   (2) Micrograms (μg) are larger than milligrams (mg).
   (3) For good health the human body needs a variety of vitamins.
   (4) Children don't need vitamins.
   (5) Vitamin A is important for night vision.

10. Which of the following most likely is an opinion, rather than a scientific fact?

   (1) Unlike human beings, many animals can produce their own vitamin C.
   (2) Vitamin deficiency can lead to many types of illnesses.
   (3) Fruits and nuts contain more essential vitamins than does candy.
   (4) Vitamin pills should be taken by everyone over the age of 30.
   (5) Both minerals and vitamins are essential for good health.

11. Computerized tomography (CT) scanning—a form of imaging done by computer analysis of X-ray pictures—is considered to be the best technique available for making images of such body parts as the bones, brain, skull, spinal cord, lungs, abdomen, and pelvis. For which two of the following injuries would you expect a CT scan to give the clearest images of affected parts?

   A. a compound fracture of the lower arm
   B. torn ligaments in the knee
   C. bruised shoulder muscles
   D. a broken leg

   (1) A and B
   (2) A and C
   (3) A and D
   (4) B and C
   (5) B and D

Question 12 refers to the diagram below.

THE HUMAN TOOTH

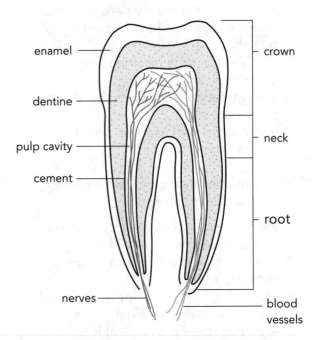

12. What is the name given to the protective outer layer of a tooth?

   (1) crown
   (2) neck
   (3) root
   (4) enamel
   (5) dentin

Questions 13 and 14 refer to the following illustrations.

TOP VIEW OF HUMAN SKULL

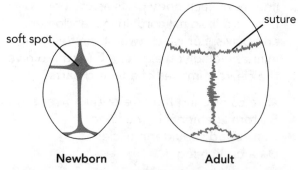

Newborn   Adult

The membrane between the separate plates of bone in the human skull slowly disappears as an infant grows. In an adult the skull pieces are completely fused together.

13. What is the best description of the soft spot on a baby's head?

  (1) a region of bone that is softer than other regions of bone on the human head
  (2) a region on the head where the several bones of the skull have not yet grown together
  (3) a hole in the skull resulting from a bone disease
  (4) a temporary birthmark on the skin that covers the skull
  (5) a tear in the skin over the skull that occurs during childbirth

14. What is the best reason to be careful <u>not</u> to touch a baby's soft spot?

  (1) No protective bone lies between the soft spot and the baby's brain.
  (2) The skin over the soft spot is very sensitive to pain.
  (3) The soft spot is more apt to get a rash than other skin on the baby's head.
  (4) A soft spot will get larger if it is touched.
  (5) A soft spot will not close as quickly if it is touched often.

Questions 15 refers to the illustration below.

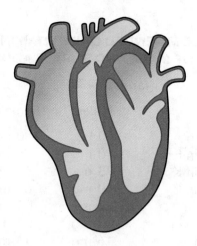

15. What is the main purpose of the organ pictured above?

  (1) to absorb oxygen from the air
  (2) to digest food
  (3) to filter blood
  (4) to pump blood
  (5) to expel carbon dioxide into the air

Questions 16–18 are based on the following information and graphs.

The three major structural tissues in the human body are fat, muscle, and bone. Fat can be either of two types: essential fat or storage fat. Essential fat is stored in bone marrow and in organs. Storage fat is the type we usually think of as excess fat, since it often adds unwanted weight to our bodies. Storage fat is fatty tissue that surrounds internal organs or is deposited beneath the skin.

The circle graphs below compare the body composition of an average-size young man with that of an average-size young woman.

BODY COMPOSITION
(percent of total weight)

Average-Sized Young Man

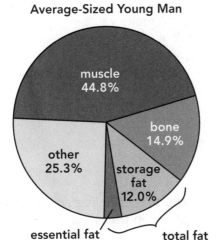

Average-Sized Young Woman

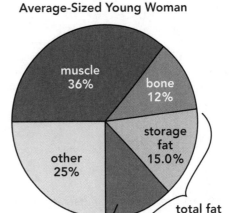

16. An average-size young man has a higher percentage of what tissue than does an average-size young woman?

    (1) total fat and bone
    (2) bone and muscle
    (3) total fat and muscle
    (4) bone only
    (5) muscle only

17. Colleen is an average-size young woman and Miguel is an average-size young man. Which statement about Colleen and Miguel is best supported by the passage and graphs?

    (1) Colleen eats more sweets than Miguel.
    (2) Colleen gets less exercise than Miguel.
    (3) Colleen has the same percent of total fat as Miguel.
    (4) Colleen has a higher percent of total fat than Miguel.
    (5) Colleen has a more difficult time losing weight than Miguel.

18. What is the most reasonable explanation for why women tend to have a higher percentage of total fat than men?

    (1) Women tend to do less physical labor than men.
    (2) Women's bodies use added fat to protect female reproductive organs.
    (3) Women have broader hips than men.
    (4) Women have larger appetites than men.
    (5) Women have different types of diets available to them than men.

Questions 19 and 20 refer to the following diagram.

ACTION OF VALVES IN VEINS

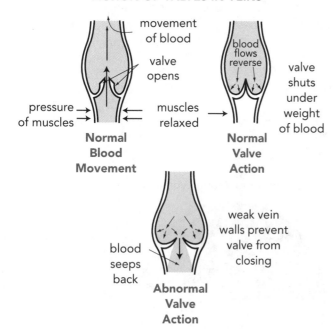

Normal Blood Movement

movement of blood

valve opens

pressure of muscles

muscles relaxed

Normal Valve Action

blood flows reverse

valve shuts under weight of blood

Abnormal Valve Action

blood seeps back

weak vein walls prevent valve from closing

19. What is the purpose of valves in veins?

   (1) to increase blood pressure
   (2) to increase the speed of blood flow
   (3) to restrict the flow of blood to a single direction
   (4) to enable blood to flow back and forth
   (5) to transfer blood from arteries to veins

20. What phrase might be used to describe the abnormal valve action indicated in the drawing?

   (1) a leaky valve
   (2) a constricted valve
   (3) a stiff valve
   (4) a permanently closed valve
   (5) a pressurized valve

Question 21 refers to the following illustration.

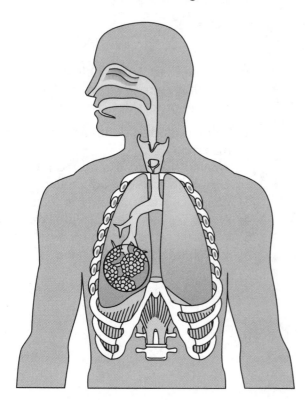

21. What function is performed by the organs shown in the drawing?

   (1) pumping blood
   (2) swallowing food
   (3) breathing air
   (4) excreting liquid waste
   (5) drinking water

Questions 22 and 23 refer to the passage below.

Temporary sunburn is the least serious of many health problems risked by sunbathers. In addition to premature wrinkles and forms of skin cancer, overexposure to sunlight can cause eye damage.

Sunlight comprises many colors of light. Most harmful to the eyes is high-energy light, mainly invisible ultraviolet light.

The *cornea* is the transparent outer surface of the eye. The cornea helps protect the inner parts of the eye by absorbing much of the Sun's ultraviolet light. However, overexposure to bright sunlight can cause *keratitis*, an itchy condition around the cornea that usually lasts a day or two. Snow blindness and welder's flash are two types of keratitis.

Behind the cornea is the *lens*. The purpose of the lens is to focus light onto the retina at the back of the eye. High levels of ultraviolet light can cause *cataracts* to form on the lens. Cataracts are cloudy areas that disrupt vision by scattering visible light.

The *retina*, the inner layer of the eye, is made up of layers of cells that change light rays to electrical signals. These signals are sent along the optic nerve to the brain. Researchers have found that prolonged exposure to ultraviolet light can scar the retina. The risk is greatest to people who work or play for long periods of time in bright sunshine.

22. What is the key point of this passage?

(1) Overexposure to sunlight is the main cause of keratitis.
(2) Sunlight comprises light of various colors and energies.
(3) Welders are more sensitive to sunlight than other people.
(4) The eye is made up of several related structures.
(5) Overexposure to sunlight can result in several types of eye damage.

23. What is the most likely health-related purpose of sunglasses?

(1) to reflect high-energy light rays
(2) to absorb high-energy light rays
(3) to reflect low-energy light rays
(4) to absorb low-energy light rays
(5) to increase light reaching the eyes

Questions 24 and 25 refer to the following diagram.

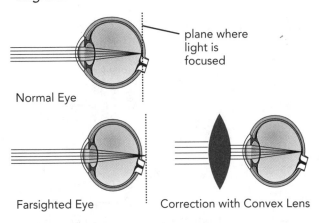

24. In a normal eye where is the focal point for light rays entering the eye?

(1) cornea
(2) iris
(3) lens
(4) retina
(5) optic nerve

25. What is a convex eyeglass lens designed to do?

(1) protect the eyes from rocks
(2) increase the focal length of light passing through the eyes
(3) decrease the focal length of light passing through the eyes
(4) decrease the amount of light entering the eyes
(5) increase the amount of light entering the eyes

Questions 26–29 refer to the following graph.

**OXYGEN CONSUMPTION RATE—
MEASURED WHILE JOGGING**

(measured at a slow-jog pace of 12 minutes per mile)

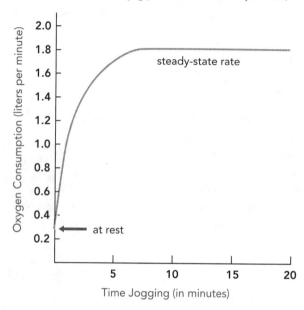

26. The average heart rate of a person at rest is about 78 beats per minute. According to the graph, about how much oxygen is consumed per minute by a healthy but nonathletic person at rest?

    (1) 0.1 liter
    (2) 0.3 liter
    (3) 0.8 liter
    (4) 1.0 liter
    (5) 1.5 liters

27. Which statement best summarizes the graph?

    (1) The steady-state rate of oxygen consumption depends on jogging speed.
    (2) Oxygen consumption increases most rapidly during the first 6 minutes of jogging.
    (3) Oxygen consumption is lower while at rest than during jogging.
    (4) Oxygen consumption rises during the first 6 minutes of jogging to a constant steady-state rate thereafter.
    (5) Different individuals have different steady-state oxygen consumption rates.

28. Suppose that after 10 minutes of jogging at a rate of 12 minutes per mile, a person suddenly speeds up to a rate of 11 minutes per mile. What will be the most likely effect on oxygen consumption?

    (1) an increase to a higher steady-state level
    (2) a decrease to a lower steady-state level
    (3) a continuation of the same steady-state level
    (4) a steadily rising rate and no new steady-state level
    (5) a steadily falling rate and no new steady-state level

29. During the 10-minute and 20-minute points shown on the graph what is a jogger's heart rate most likely doing?

    (1) slowly increasing at a steady rate
    (2) slowly decreasing at a steady rate
    (3) slowly increasing and then decreasing
    (4) slowly decreasing and then increasing
    (5) remaining at a constant level

Questions 30 and 31 refer to the following illustrations.

COMPARISON OF HUMAN AND BABOON JAWS

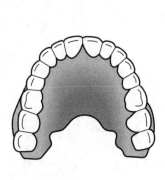

Human

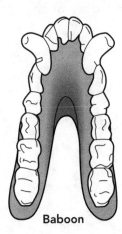

Baboon

30. What feature of a human jaw is similar to that of a baboon?

(1) length of the jaw
(2) width of the jaw
(3) number of teeth in the jaw
(4) shape of teeth in the jaw
(5) shape of the jawbone

31. Differences between the jaw of a human being and the jaw of a baboon most likely are related to differences in what activity?

(1) intelligence
(2) life span
(3) breathing
(4) diet
(5) communication

32. Lung tissue is damaged when exposed to smoke over a prolonged period of time. When lung tissue is damaged, less oxygen enters the bloodstream from the lungs.

Suppose that a smoker and a nonsmoker begin an exercise program. Compared with the nonsmoker, what would the smoker most likely experience during mild exercise?

(1) a slower breathing rate
(2) less need to rest frequently
(3) muscles tiring more slowly
(4) a smaller amount of perspiration
(5) a more rapidly beating heart

33. Which fact is <u>least</u> relevant to the relationship between the consumption of alcohol and good health?

(1) Alcohol affects some body organs more than others.
(2) People who drink a lot tend to ignore their nutritional needs.
(3) When pregnant women drink alcohol, the developing fetus can be harmed.
(4) Seventy percent of drivers killed in one-car accidents had been drinking alcohol shortly before the accident.
(5) One can of beer contains the same amount of alcohol as one glass of wine.

Question 34 is based on the following passage and illustration.

The following illustration shows how genes are passed from generation to generation.

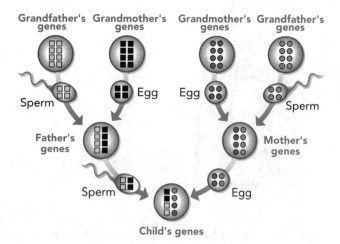

Questions 35 and 36 refer to the following illustration.

TASTE BUDS ON THE TONGUE

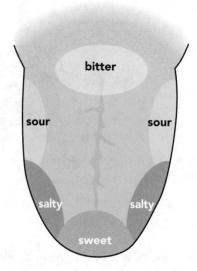

34. Which of the following is <u>not</u> a conclusion that you can draw from the illustration above?

   (1) The gene inherited from the male sperm is dominant over the gene inherited from the female egg.
   (2) The number of genes in a sperm cell or egg cell is half the number of genes in the regular parent cells.
   (3) Personal traits are based not only on the traits of parents but also on the traits of grandparents.
   (4) Combinations of genes intermix from one generation to the next.
   (5) A child inherits genes from its parents while the child's parents inherited genes from the child's grandparents.

35. Which of the following is the tip of your tongue most sensitive to tasting?

   (1) vinegar
   (2) syrup
   (3) lemon juice
   (4) salt
   (5) walnut

36. What is a reasonable conclusion to draw from looking at the illustration?

   (1) The most important taste is sweetness because taste buds for it are located on the front of the tongue.
   (2) Foods that have a sweet taste are more nutritious than foods that are not sweet.
   (3) The least important taste is bitterness because taste buds for it are located on the back of the tongue.
   (4) Human beings are sensitive to four basic tastes.
   (5) Taste buds are located on the edge of the tongue because the middle part does not have nerve endings.

Questions 37 and 38 are based on the following passage.

Two million Americans suffer from some form of epilepsy. Epilepsy is a disorder of nerve cells in the brain. Normally, nerve cells in the brain produce electrical signals that flow through the nervous system and activate body muscles. However, during an epileptic seizure, these cells release abnormal bursts of electrical energy that the brain cannot control.

In the most severe type of epilepsy, people may lose consciousness, fall, and shake. This type of seizure often lasts several minutes. In a milder form, people may lose awareness of their surroundings for a few seconds but do not fall or lose consciousness.

Scientists do not understand the causes of epilepsy, but they do know that it cannot be spread from one person to another. Moreover, doctors are now able to treat epilepsy with drugs that either reduce the frequency of seizures or prevent them entirely. Most people who have epilepsy can now lead normal lives.

37. What is the best description of an epileptic seizure?

(1) a temporary loss of body control due to an electrical disturbance in the brain
(2) a permanent loss of body control due to an electrical disturbance in the brain
(3) an electrical disturbance in the brain caused by a loss of body control
(4) a permanent mental condition caused by severe bodily injury
(5) an injury to the body caused by a fall that occurs because of a brain malfunction

38. Which phrase is <u>least</u> descriptive of epilepsy?

(1) electrical disturbance in the brain
(2) characterized by loss of body control
(3) a contagious disease
(4) often accompanied by a seizure
(5) often controllable with medicine

Questions 39 and 40 refer to the following illustrations.

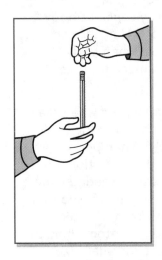 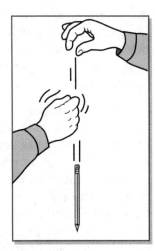

39. What is the experiment above most likely designed to test?

(1) hand strength
(2) strength of gravity
(3) slipperiness of skin oil
(4) weight of a pencil
(5) reaction time

40. Which activity would most likely influence a person's ability to perform well on this test?

(1) eating vegetables
(2) drinking alcohol
(3) playing a computer game
(4) watching a movie
(5) talking on the phone

Questions 41–44 refer to the passage below.

Under normal conditions, human beings maintain a relatively constant body temperature—about 98.6°F. Humans can become very ill or even die when body temperature changes too much from its normal level.

Heatstroke (sunstroke) occurs when the temperature-regulating system of the body ceases to work effectively. When people are exposed to or undergo heavy exertion in extreme heat, they may stop sweating—the body's natural way of cooling itself. The skin becomes hot and dry, and body temperature rises above normal. Other symptoms may include irregular heartbeat and shallow, irregular breathing. The victim usually loses consciousness.

Because the high body temperatures of heatstroke can cause brain damage and death, heatstroke should be treated immediately. Standard treatment is to quickly reduce body temperature by applying cold compresses to the victim's body and ice packs to the neck. If possible, the victim should be placed in a bathtub full of cold water. Only when body temperature is down to 102°F should the cooling procedures be stopped.

Heat exhaustion is less serious than heatstroke. With heat exhaustion, a victim becomes weak and dizzy after exposure to or heavy exertion in high temperature and high humidity. Other symptoms include confusion, a great amount of sweating, and a body temperature <u>below</u> normal. Proper treatment includes moving the victim to a cooler location, but keeping the victim warm until body temperature rises to normal.

To help avoid heatstroke and heat exhaustion, people exposed to hot and possibly humid conditions should drink plenty of water and take frequent rest breaks in order to cool off.

41. Which of the following is <u>not</u> a symptom of heat exhaustion?

    (1) mental confusion
    (2) body weakness
    (3) dizziness
    (4) profuse sweating
    (5) elevated temperature

42. What is the main purpose of applying ice packs to the neck of a heatstroke victim?

    (1) to slow the flow of blood to the brain
    (2) to lower body temperature by inducing shivering
    (3) to cool the shoulder muscles
    (4) to cool the blood flowing to the brain
    (5) to cool the nervous system

43. To help prevent heat exhaustion during summer, which is the <u>least</u> important thing for an exercise club to maintain in good working order?

    (1) exercise bicycles
    (2) air conditioner
    (3) bathroom sinks
    (4) ventilation system
    (5) drinking fountains

44. A doctor comes to the assistance of a middle-aged woman who has collapsed while working outdoors on a hot afternoon. Which of the following is the <u>least</u> important clue that the doctor would need to know?

    (1) heart rate
    (2) body temperature
    (3) blood pressure
    (4) air temperature
    (5) breathing rate

Answers are on page 111.

-Directions: Choose the <u>one best answer</u> to each question.

Questions 1 and 2 refer to the following diagram.

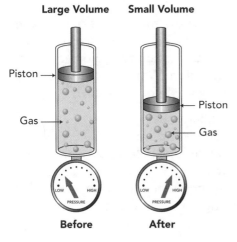

Large Volume   **Small Volume**

Piston

Gas

Piston

Gas

Before   **After**

As the piston is moved down in the cylinder, the temperature of the gas is kept constant.

1. What conclusion can be drawn from this experiment?

    (1) Gas temperature increases as gas volume decreases.
    (2) Gas temperature decreases as gas volume increases.
    (3) Gas pressure increases as gas temperature increases.
    (4) Gas pressure increases as gas volume decreases.
    (5) Gas pressure decreases as gas volume decreases.

2. For the design of which product would data from this experiment be most useful?

    (1) parachute
    (2) steam engine
    (3) snow skis
    (4) airplane wings
    (5) bicycle tires

Questions 3 and 4 refer to the following passage.

Elements are classified as metals, nonmetals, or metalloids, according to their properties.

*   *Metals*—elements that are good conductors of electricity and are malleable (can be softened by striking with a hammer), shiny, and ductile (can be stretched)
*   *Nonmetals*—elements that are not good conductors of electricity, and are not shiny, malleable, or ductile
*   *Metalloids*—elements that have some of the properties of metals and some of the properties of nonmetals

3. The element sulfur is not shiny and will not conduct electricity. Sulfur is not malleable and is not ductile. What is the best way to classify sulfur?

    (1) metal
    (2) nonmetal
    (3) metalloid
    (4) gaseous metal
    (5) liquid metal

4. Tellurium is a shiny element that is very brittle and can be ground easily into a powder. What is the best way to classify tellurium?

    (1) gaseous nonmetal
    (2) liquid nonmetal
    (3) metalloid
    (4) metal
    (5) liquid metal

Questions 5 and 6 refer to the illustration below.

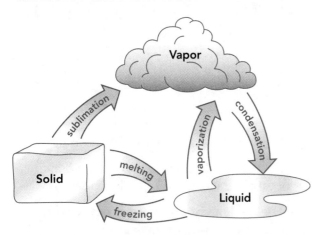

5. What change of phase takes place during sublimation?

   (1) from gas to liquid
   (2) from gas to solid
   (3) from liquid to solid
   (4) from solid to gas
   (5) from solid to liquid

6. Which of the following is an example of condensation?

   (1) the heating of iron until it is a liquid
   (2) the formation of dew
   (3) the making of instant coffee
   (4) the making of ice
   (5) the heating of milk until vapor forms

7. Ozone (O_3) is produced in the upper atmosphere when oxygen gas (O_2) absorbs high-energy sunlight. Remembering that the number of oxygen atoms must be the same on each side of the equation, which equation correctly describes this reaction?

   (1) O_2 + energy $\rightarrow$ O_2
   (2) O_2 + energy $\rightarrow$ $2O_2$
   (3) $2O_2$ + energy $\rightarrow$ $3O_2$
   (4) $3O_2$ + energy $\rightarrow$ $2O_3$
   (5) $3O_2$ + energy $\rightarrow$ $3O_3$

8. *Diffusion* is the movement of the atoms or molecules of a substance from regions of high concentration to regions of lower concentration. Which of the following is <u>not</u> an example of diffusion?

   (1) freezing meat
   (2) evaporating water
   (3) soaking up spilled pop with a paper towel
   (4) putting cream in coffee
   (5) using an air freshener

Questions 9 and 10 refer to the following diagram.

ELECTROLYSIS

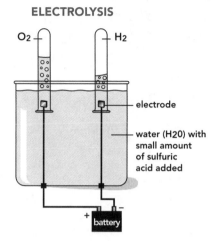

9. What is the best definition of *electrolysis*?

   (1) the use of an electric current to cause bubbles to form in a liquid
   (2) the use of a battery to cause an electric current to flow through a liquid
   (3) the use of an electric current to break down a substance into its component atoms or molecules
   (4) the use of a battery to create oxygen gas
   (5) the use of water to create gas by causing electrodes to rust

10. Which of the following would increase the rate at which oxygen and hydrogen gases form during electrolysis?

   (1) replacing the battery with one of higher voltage
   (2) replacing the battery with one of lower voltage
   (3) disconnecting the battery
   (4) increasing the amount of water in the jar
   (5) increasing the size of the inverted glass tubes

Question 11 refers to the following passage and graph.

Oil flows at a different speed in each section of the pipe shown below. According to the Principle of Continuity of Fluid Flow, the speed of the oil is greatest in the section of pipe with the smallest diameter. The speed of oil is least in the section of pipe with the largest diameter.

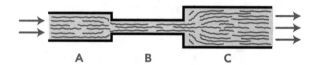

11. Which listing correctly identifies from slowest to fastest the relative speed of the oil flowing through the three sections of pipe?

   (1) B, A, C
   (2) B, C, A
   (3) C, A, B
   (4) C, B, A
   (5) A, C, B

Questions 12–14 refer to the passage below.

Acid rain contains high levels of sulfuric acid and/or nitric acid. These acids form when sulfur dioxide and nitric oxide gases react chemically with oxygen and water in the atmosphere.

• Sulfur dioxide gas is produced when electric and industrial plants burn coal and oil fuels that contain sulfur.

• Nitric oxide is produced by car engines.

Acid rain can destroy fish and other water life, and it poses a threat to forests and croplands. It can also affect people, causing serious injury to the moist surfaces of the eyes and to the membranes in the lungs.

12. What are two gaseous pollutants involved in the production of acid rain?

   (1) oxygen and water
   (2) coal and oil fuels
   (3) sulfuric acid and nitric acid
   (4) sulfur dioxide and nitric oxide
   (5) electrical and industrial plants

13. Which activity would most likely contribute to acid rain?

   (1) burning oil in a home furnace
   (2) boiling water on a stove
   (3) popping a helium-filled balloon
   (4) using an electric current to break down water into oxygen and hydrogen gas
   (5) making a campfire with old branches

14. Which is the <u>least</u> important rule of thumb for people living in a city that has a serious acid rain problem?

   (1) Stay indoors when acid rain is falling.
   (2) Wear extra warm clothes if you must be outside when acid rain is falling.
   (3) Don't drive your car when acid rain is falling.
   (4) If you get wet in acid rain, shower and wash your clothes as soon as possible.
   (5) Wear a raincoat and use an umbrella when acid rain is falling.

15. Iron (Fe) can combine with oxygen (O) to produce any one of three different iron oxides: FeO, Fe_2O_3, and Fe_3O_4.

Of the three oxides FeO is the most unstable. If FeO is exposed to air, it forms Fe_2O_3. Which equation correctly describes this reaction?

   (1) $FeO + O_2 \rightarrow Fe_2O_3$
   (2) $FeO + O_2 \rightarrow 2Fe_2O_3$
   (3) $2FeO + O_2 \rightarrow Fe_2O_3$
   (4) $4FeO + O_2 \rightarrow Fe_2O_3$
   (5) $4FeO + O_2 \rightarrow 2Fe_2O_3$

Question 16 refers to the chart below.

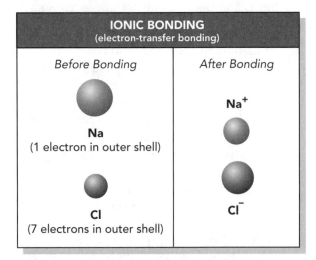

IONIC BONDING
(electron-transfer bonding)

Before Bonding

Na
(1 electron in outer shell)

Cl
(7 electrons in outer shell)

After Bonding

Na⁺

Cl⁻

16. What happens during the bonding of a sodium atom (Na) to a chlorine atom (Cl)?

   (1) A proton (+) transfers from the sodium atom to the chlorine atom.
   (2) A proton (+) transfers from the chlorine atom to the sodium atom.
   (3) An electron (–) transfers from the sodium atom to the chlorine atom.
   (4) An electron (–) transfers from the chlorine atom to the sodium atom.
   (5) Two electrons are shared by the sodium and chlorine atoms.

Questions 17–20 refer to the following chart.

Property of Gas	Description
loose molecular structure	A gas consists of atoms or molecules that rapidly move in straight lines until they collide with each other or with the walls of a container. The distances between these particles are very large compared with the size of the particles. In liquids and solids the particles are much closer together.
temperature-dependent energy	Each atom or molecule of gas has kinetic energy (energy of motion) that increases as the temperature of the gas increases. As a gas becomes hotter, its molecules gain kinetic energy.
equal pressure	A gas exerts equal pressure (force per unit area) on each section of wall of any container to which it is confined.
diffusion	When two gases are brought into contact, they will freely mix—the molecules of each gas intermingling with the molecules of the other gas.
effusion	Gas atoms or molecules will pass through a small opening from an area of higher pressure to an area of lower pressure.

17. A basketball keeps its round shape as more air is added. Which property accounts for the fact that each part of the ball feels equally hard as it is being filled?

   (1) loose molecular structure
   (2) temperature-dependent energy
   (3) equal pressure
   (4) diffusion
   (5) effusion

18. A small child is holding a fully inflated balloon when suddenly she releases it. As air escapes through the mouth of the balloon, the balloon flies around the room. What property accounts for what is happening?

   (1) loose molecular structure
   (2) temperature-dependent energy
   (3) equal pressure
   (4) diffusion
   (5) effusion

19. Unlike solids and liquids, gases can be compressed to a small fraction of their original volume. Fire extinguishers and other containers of compressed gas depend on this property. What property explains the reason that gases are so compressible?

   (1) loose molecular structure
   (2) temperature-dependent energy
   (3) equal pressure
   (4) diffusion
   (5) effusion

20. Just before leaving for work, Allison sprayed herself lightly with perfume. Within a few seconds, the perfume's odor had spread throughout the room. What property is exemplified by the gaseous vapor of this perfume?

   (1) loose molecular structure
   (2) temperature-dependent energy
   (3) equal pressure
   (4) diffusion
   (5) effusion

Questions 21–23 refer to the information below.

Viscosity is a measure of how much a fluid resists flowing. Viscosity is related to the "thickness" of a liquid, and it changes with temperature. At room temperature the viscosity of syrup is much greater than the viscosity of water; the syrup pours more slowly than water. However, if syrup is heated, it pours almost as easily as water.

Viscosity is determined by the strength of chemical bonds that hold molecules of a liquid together. These bonds form a resistance that retards the movement of molecules past one another. Heating causes molecules to move more quickly and weakens the bonds that hold them together.

21. At room temperature which of the following liquids has the strongest molecular bonds?

    (1) orange juice
    (2) diet cola
    (3) milk
    (4) honey
    (5) water

22. In winter a car's engine is slower to turn over than in summer. What is the most likely reason for this?

    (1) Engine oil thickens in cold weather.
    (2) Gasoline flows more slowly in cold weather.
    (3) Radiator water freezes in cold weather.
    (4) Electric current flows more slowly in cold weather.
    (5) An engine is more likely to overheat in winter than in summer.

23. What is one way to increase the viscosity of a liquid?

    (1) Heat it until it's boiling.
    (2) Shake it until it's full of bubbles.
    (3) Cool it until it's frozen.
    (4) Dilute it with water.
    (5) Cool it but do not freeze it.

Questions 24 and 25 refer to the passage below.

Chemistry is the study of the structure, composition, and properties of matter. In a classroom discussion students made the following statements about the study of chemistry.

    A. Chemistry is the most difficult science to learn.
    B. There are many practical applications of chemistry.
    C. Much of the knowledge gained from chemistry is useful in geology.

24. Which statement(s) above is most likely fact and not opinion?

    (1) A only
    (2) B only
    (3) C only
    (4) Both A and B
    (5) Both B and C

25. Which statement(s) above is most likely opinion and not fact?

    (1) A only
    (2) B only
    (3) C only
    (4) Both A and B
    (5) Both B and C

Questions 26 and 27 refer to the graph below.

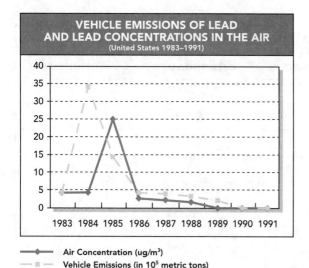

VEHICLE EMISSIONS OF LEAD AND LEAD CONCENTRATIONS IN THE AIR
(United States 1983–1991)

——— Air Concentration (ug/m³)
– – ■ – – Vehicle Emissions (in 10⁵ metric tons)

26. What is the best summary of the information presented on the graph?

    (1) Between 1983 and 1991 lead emissions decreased greatly.
    (2) Between 1983 and 1991 lead pollution in the air decreased at about the same rate as vehicle emissions of lead.
    (3) Between 1983 and 1991 lead poisoning in children decreased greatly.
    (4) Between 1983 and 1991 new pollution-control devices were installed in cars.
    (5) Between 1983 and 1991 lead pollution in the air decreased greatly.

27. What else would you need to know to be able to conclude that between 1983 and 1991 lead pollution in the air decreased mainly because of a decrease in lead emissions by vehicles?

    A. Whether lead used in gasoline results in lead pollution of the air
    B. The number of people who switched from leaded to unleaded gasoline between 1983 and 1991
    C. How lead pollution from other sources varied between 1983 and 1991

    (1) A only
    (2) B only
    (3) C only
    (4) Both A and C
    (5) Both B and C

Questions 28 and 29 refer to the following passage.

When chemical or physical reactions occur, energy is either given off or absorbed.

• In an *exothermic reaction* energy is given off. An example is a forest fire, which produces heat and light energy.

• In an *endothermic reaction* energy is absorbed. Photosynthesis—the absorption of sunlight by a plant in order to produce sugar—is an example.

28. Which of the following is an endothermic reaction?

    (1) water turning into ice
    (2) cookies baking in an oven
    (3) a flashlight producing a beam of light
    (4) a firecracker exploding
    (5) a candle burning

29. If hydrogen (H) and oxygen (O) gases freely mix, they combine in an explosive reaction to form water (H_2O). Heat, sound, and light are created during this reaction.

    What type of reaction occurs when hydrogen and oxygen mix?

    (1) endothermic, because water has more chemical energy than separated hydrogen and oxygen gases
    (2) endothermic, because water has less chemical energy than separated hydrogen and oxygen gases
    (3) exothermic, because water has more chemical energy than separated hydrogen and oxygen gases
    (4) exothermic, because water has less chemical energy than separated hydrogen and oxygen gases
    (5) neither endothermic nor exothermic

Questions 30 and 31 refer to the table below.

COMPOSITION OF SEVERAL TYPES OF COAL				
	Lignite	*Subbituminous*	*Bituminous*	*Anthracite*
Carbon, %	37	51	75	86
Hydrogen, %	7	6	5	3
Oxygen, %	48	35	9	4
Moisture, %	43	26	3	2
Nitrogen, %	0.7	1.0	1.5	0.9
Sulfur, %	0.6	0.3	0.9	0.6
Energy Value, (kcal/kg)	3,488	4,784	7,441	7,683

30. What element is most responsible for the energy value of coal?

(1) carbon
(2) hydrogen
(3) oxygen
(4) nitrogen
(5) sulfur

31. For what purpose is the information presented in the table most likely used?

(1) the abundance of each type of coal
(2) the region in which each type of coal is found
(3) the selling price of each type of coal
(4) the cost of digging each type of coal
(5) the estimated total reserves of all types of coal

32. An *emulsion* is a suspension of small droplets of one liquid in a second liquid with which the first liquid will not mix freely. An example is tiny droplets of milk fat suspended in cream. Emulsions are common in food products and usually are typical of foods having a thick or creamy consistency.

Which of the following products is <u>not</u> an emulsion?

(1) mayonnaise
(2) yogurt
(3) spaghetti sauce
(4) apple cider
(5) mustard

33. The pressure of a gas in a closed container can be increased by increasing the temperature of the gas or by adding more gas to the container.

Which of the following does <u>not</u> utilize these properties of gas pressure?

(1) an automobile tire
(2) a sail on a sailboat
(3) a pressure cooker for use in home cooking
(4) a basketball
(5) a pressure-relief valve on a car radiator

Questions 34–37 refer to the following illustration.

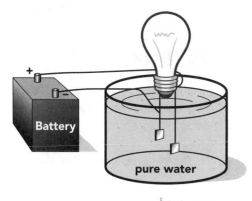

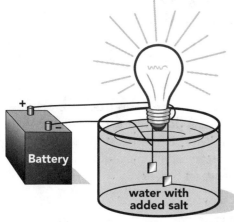

34. What is the key point made in the illustration?

  (1) Salt dissolves in pure water only when an electric current is flowing.
  (2) Dissolved salt changes pure water into a conductor of electricity.
  (3) Dissolved salt forms positive and negative ions in pure water.
  (4) Dissolved salt changes pure water into a nonconductor of electricity.
  (5) Electric appliances should never be placed in salt water.

35. What can the equipment shown in the illustration be used to test?

  (1) the strength of a battery
  (2) the presence of salt in a sample of water
  (3) the temperature of a sample of water
  (4) the weight of a sample of water
  (5) the wattage of a lightbulb

36. Which of the following liquids most likely has electrical properties similar to those of salt water?

  (1) a water and antifreeze solution used in a car radiator
  (2) distilled water used in a steam iron
  (3) a water and sulfuric acid solution used in a car battery
  (4) a water and sugar solution used in syrup
  (5) a water and dye solution used to dye clothes

37. As shown by the graph below, the amount of current that a quart of water will conduct increases up to a certain point as more salt is added.

SOLUBILITY AND ELECTRIC CURRENT

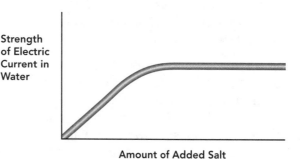

What reasonable conclusion(s) can you draw from the graph?

A. The strength of an electric current depends on the amount of salt dissolved in water.
B. For some amount of added salt, a quart of water becomes saturated and no more salt will dissolve.
C. The amount of salt that will dissolve in a quart of water depends on the temperature of the water.

  (1) A only
  (2) B only
  (3) C only
  (4) Both A and B
  (5) Both A and C

Questions 38–41 refer to the following passage and table.

An *acid* is a substance that releases hydrogen when mixed with water. An acid is sour to the taste. *Organic acids* are not harmful and are found in foods. Citric acid is found in oranges, grapefruit, and lemons. Acetic acid is found in vinegar.

Inorganic acids, on the other hand, are both strong and dangerous and can cause severe skin burns. Sulfuric acid is used in batteries. Hydrochloric acid is used as a metal cleanser.

A *base* is a substance that combines with an acid and neutralizes its effect. Together, an acid and a base form a salt and water. Baking soda, a weak base, can be used both as a safe cleanser and as a mild stomach antacid. Stronger bases, such as ammonia and bleach, are used as high-strength cleansers and can be harmful if touched or ingested.

Chemists measure the strength of an acid or a base by use of a pH scale. Values on a pH scale range from 0 to 14, with the value 7 representing neutrality. Acids have a pH value of less than 7, while bases have a pH value greater than 7. Pure water is neither an acid nor a base and has a pH value of 7. Common pH values are shown below.

pH VALUES OF COMMON SUBSTANCES		
battery acid	0.2	
normal stomach acid	2.0	
soft drinks	3.0	*more*
orange juice	3.5	*acidic*
banana	4.6	
bread	5.5	
potatoes	5.8	
rainwater	6.2	
milk	6.5	
pure water	7.0	*neutral*
eggs	7.8	
hair shampoo	8.7	*more*
household bleach	12.8	*basic*

38. If a small amount of sodium bicarbonate (baking soda) is added to a glass of orange juice, how would the taste of the juice change?

    (1) The juice would taste more acidic.
    (2) The juice would taste the same.
    (3) The juice would taste very basic.
    (4) The juice would taste less acidic.
    (5) The juice would taste colder.

39. What is the most likely pH value of acid rain?

    (1) greater than 12.8
    (2) between 7 and 12.8
    (3) exactly 7
    (4) exactly 6.2
    (5) less than 6

40. The needles of evergreen trees are acidic, and lawn grass does not grow well in acidic soil. What is the most likely reason that gardeners often place lime on lawns near evergreen trees?

    (1) Lime is a base.
    (2) Lime is an acid.
    (3) Lime is an element.
    (4) Lime is a salt.
    (5) Lime is a neutral substance.

41. Taken together, which of the following will best settle an acidic stomach?

    (1) a soft drink and a piece of bread
    (2) a glass of apple juice and a banana
    (3) a glass of water and a hard-boiled egg
    (4) a glass of orange juice and a banana
    (5) a glass of milk and a potato

Question 42 refers to the following illustration.

42. What is the most likely capacity of the smaller glass beaker shown above?

(1) 2.5 milliliters
(2) 25 milliliters
(3) 250 milliliters
(4) 2.5 liters
(5) 250 liters

Questions 43 and 44 refer to the following passage.

Many exothermic reactions—reactions that give off heat—require that energy be supplied before the reactions can start. This energy is called *activation energy*. As an example, heat is used to cause a firecracker to explode. The explosion itself is an exothermic reaction that gives off energy as heat, sound, and light.

43. What phrase best describes activation energy?

(1) energy shared by reacting substances
(2) energy required to complete a reaction
(3) energy given off during a reaction
(4) energy given off before a reaction starts
(5) energy needed to start a reaction

44. Which example below is an exothermic reaction that needs activation energy?

(1) water cooling in a refrigerator
(2) glue drying in a warm room
(3) a candle burning
(4) a fever, occurring during illness
(5) bread being toasted

Questions 45 and 46 refer to the following passage and graph.

Density is a defining property of matter and is measured as the amount of matter per unit volume. If two objects have the same volume, the object with the higher density has more matter and is heavier. The graph shows the density in grams per cubic centimeter of several common substances.

DENSITY OF SEVERAL ELEMENTS

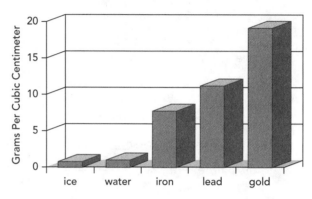

45. About how many cubic centimeters of iron are equal in weight to one cubic centimeter of gold?

(1) 2.4
(2) 1.9
(3) 1.5
(4) 1.1
(5) 0.7

46. Which of the following statements can you infer to be true?

(1) On average, H_2O molecules are farther apart in ice than in water vapor.
(2) On average, H_2O molecules are farther apart in water than in ice.
(3) On average, H_2O molecules are the same distance apart in water and in ice.
(4) On average, H_2O molecules are closer together in water vapor than in water.
(5) On average, H_2O molecules are closer together in water than in ice.

Questions 47 and 48 refer to the following information.

Below are three examples of chemical reactions that go on around us much of the time.

Car Battery

$$PbO_2 + 2H_2SO_4 + Pb \rightarrow 2PbSO_4 + 2H_2O$$

Lead Oxide   Sulfuric Acid   Lead   Lead Sulfate   Water

Car Engine Combustion

$$2C_8H_{16} + 25O_2 \rightarrow 16CO_2 + 18H_2O$$

Gasoline   Oxygen   Carbon Dioxide   Water

Atmospheric Reactions

Ozone Creation

$$\text{ultraviolet light} + 3O_2 \rightarrow 2O_3$$

Oxygen   Ozone

Acid Rain Creation

$$SO_3 + H_2O \rightarrow H_2SO_4$$

Sulfur Trioxide   Water   Sulfuric Acid Vapor

47. What gases, shown in the reaction equations above, are harmful to human health?

(1) water vapor and sulfuric acid vapor
(2) carbon dioxide and ozone
(3) water vapor and carbon dioxide
(4) ozone and sulfuric acid vapor
(5) carbon dioxide and ozone

48. When a battery is recharged, what happens to the lead that is in the battery as lead sulfate?

(1) It is converted to pure lead and water.
(2) It is converted to pure lead and lead oxide.
(3) It is entirely converted to pure lead.
(4) It is entirely converted to sulfuric acid.
(5) It is converted to sulfuric acid and pure lead.

Questions 49 and 50 refer to the following passage and illustration.

As part of an experiment, the container below is divided into two parts by a barrier. Pure water is placed on the left side of the barrier, and liquid food coloring on the right. The purpose of the experiment is to demonstrate the process of *osmosis*—the movement of molecules across a barrier (or membrane) from a region of higher concentration to a region of lower concentration.

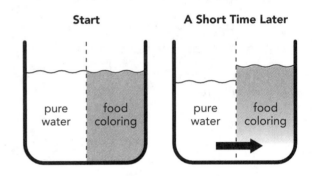

49. Which of the following can you infer to be true?

(1) Molecules of water are closer together than molecules of food coloring.
(2) Molecules of water cannot easily, if at all, pass through the barrier.
(3) Molecules of food coloring cannot easily, if at all, pass through the barrier.
(4) Molecules of food coloring can easily pass through the barrier.
(5) Molecules of food coloring do not attempt to pass through the barrier.

50. Suppose that molecules of food coloring can move across the barrier, although much more slowly than molecules of water. Where will the water eventually be if the container is covered so that no liquid evaporates?

(1) all on the right side of the barrier
(2) all on the left side of the barrier
(3) one-third on the right side of the barrier and two-thirds on the left side
(4) one-half on each side of the barrier
(5) one-third on the left side of the barrier and two-thirds on the right side

Answers are on page 112.

Directions: Choose the <u>one best answer</u> to each question.

Questions 1–5 refer to the following information.

Energy can appear in many forms and can change from one form to another.

Six forms of energy are listed below.

- *Electricity*—energy carried by a moving stream of electrons

- *Electromagnetic radiation*—energy-carrying light waves that can travel through a vacuum

- *Sound*—vibrational energy that travels through air, water, and other substances

- *Heat*—energy associated with the random motion of molecules

- *Chemical energy*—energy released when substances undergo chemical changes

- *Mechanical energy*—energy of motion (kinetic energy) associated with everyday machines such as a moving bicycle

1. A liquid evaporates when fast-moving molecules on the surface break the bonds that hold them and escape into the air. What type of energy do the escaping molecules display?

   (1) electricity
   (2) sound
   (3) heat
   (4) chemical energy
   (5) mechanical energy

2. Mileage is a measure of a car's efficiency at converting gasoline to the car's movement. In this process what type of energy results from the conversion of the chemical energy of gasoline?

   (1) electricity
   (2) electromagnetic radiation
   (3) heat
   (4) chemical energy
   (5) mechanical energy

3. Sunlight, unlike other forms of energy, travels easily through the vacuum of space. What name is given to this type of energy?

   (1) electricity
   (2) electromagnetic radiation
   (3) sound
   (4) heat
   (5) chemical energy

4. In a car battery electricity is created when lead at one electrode reacts with a solution of sulfuric acid and produces lead sulfate at a second electrode. What type of energy does a car battery use?

   (1) electromagnetic radiation
   (2) sound
   (3) heat
   (4) chemical energy
   (5) mechanical energy

5. In a superconductor no energy is lost as electrons move from one point to another. What type of energy does a superconductor carry?

   (1) electricity
   (2) electromagnetic radiation
   (3) heat
   (4) sound
   (5) chemical energy

Questions 6 and 7 refer to the chart below.

Question 8 refers to the diagram below.

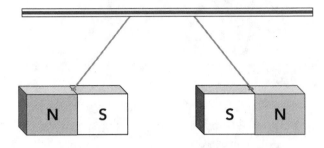

**AVERAGE COSTS OF OPERATING
SELECTED CONSUMER PRODUCTS
FOR 1 HOUR***

Air conditioner	8¢
Washing machine	4¢
Refrigerator	2¢
Dishwasher	9¢
Microwave oven	11¢
Range oven	48¢
27-inch TV	2¢
Computer	2¢
Stereo	0.9¢
Hair dryer	10¢

* Based on an electric power cost of 8.7¢ per kilowatt-hour

6. The cost of operating an electric consumer product is most related to which factor?

   (1) weight of the product
   (2) amount of space inside the product
   (3) amount of noise made by the product
   (4) amount of heat produced by the product
   (5) size of the product

7. What would you need to know to conclude that a family spends more money for electricity watching television each month than it spends washing clothes?

   A. The number of days each month that the TV is watched
   B. The number of hours each month that the TV is watched
   C. The number of times during the day that the washing machine is being used
   D. The number of hours each month the washing machine is being used

   (1) A and B
   (2) A and C
   (3) B and C
   (4) B and D
   (5) C and D

8. Each of two magnets is hanging by a string. The magnets have two forces acting on them:

   • gravitational force pulling them together
   • magnetic force pushing them apart

   Which of the following statements is true about the magnets?

   (1) The magnetic force is equal in strength to the gravitational force.
   (2) The magnetic force is stronger than the gravitational force.
   (3) The magnetic force is not as strong as the gravitational force.
   (4) The magnetic force acts only on the south (S) poles of the magnets.
   (5) The gravitational force is pushing the magnets apart.

9. Buoyancy is the tendency of an object to float in a liquid or rise in a gas. Of the following objects, which is the only one that is <u>not</u> specifically designed to make use of the buoyancy principle?

   (1) submarine
   (2) hot-air balloon
   (3) life preserver
   (4) parachute
   (5) canoe

10. At the distances shown, the gravitational force is greatest between which pair of objects?

(1) 6 kg   6 kg

(2) 6 kg   6 kg

(3) 8 kg   8 kg

(4) 8 kg   8 kg

(5) 6 kg   8 kg

11. Of the following objects, which is the only one that is <u>not</u> specifically designed to make use of the reflection property of light?

  (1) a mirror
  (2) a camera lens
  (3) high-gloss paint
  (4) furniture wax
  (5) sandpaper

12. Electrons are negatively charged particles that orbit the nucleus of an atom. The *gravitational force* between the electrons tries to pull the electrons together. However, the *electrostatic force* between the electrons succeeds in keeping the electrons widely separated from each other.

Which can you conclude about the two forces acting on orbiting electrons?

  (1) The attractive electrostatic force is weaker than the gravitational force.
  (2) The repulsive electrostatic force is weaker than the gravitational force.
  (3) The attractive electrostatic force is stronger than the gravitational force.
  (4) The repulsive electrostatic force is stronger than the gravitational force.
  (5) The repulsive electrostatic force and the gravitational force are of equal strength.

13. For a teeter-totter to balance, the product of the weight and the distance must be the same on each side of the fulcrum.

Which drawing below correctly shows a balanced teeter-totter?

(1)
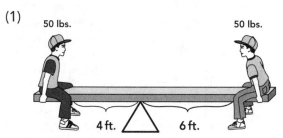

(2)
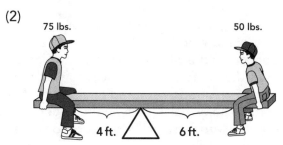

(3)
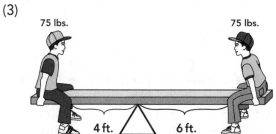

(4)
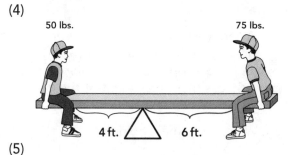

(5)
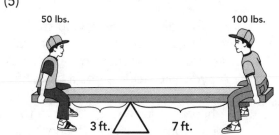

Questions 14–17 refer to the diagram and passage below.

OPERATION OF A HYDRAULIC JACK

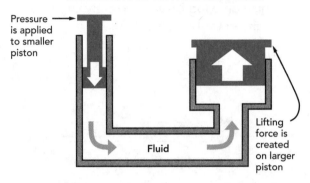

Pressure is applied to smaller piston

Lifting force is created on larger piston

Fluid

A hydraulic jack, shown above, consists of a cylinder and two pistons. When a force is applied to the smaller piston (shown on the left), that piston pushes down against the fluid. Pressure transmitted through the fluid creates an upward force on the larger piston. This upward force is in direct proportion to the area of the larger piston.

For example, suppose that the larger piston has twenty times the area of the smaller piston. A force of 1 pound pushing down on the smaller piston results in a 20-pound lifting force on the top of the larger piston. A 5-pound force on the smaller piston results in a 100-pound lifting force on the top of the larger piston.

When the larger piston of a hydraulic jack is placed under a car, a person can lift the car by applying a force on the smaller piston that is much less than the weight of the car.

14. What is the main point made in the passage?

(1) A hydraulic jack uses two pistons.
(2) A hydraulic jack enables you to raise a large piston by pushing a small piston.
(3) A hydraulic jack creates a large lifting force from a smaller pushing force.
(4) A hydraulic jack uses a fluid to transmit force from one piston to another.
(5) A hydraulic jack uses metal pistons.

15. Suppose that the larger piston of the jack shown has an area 100 times larger than that of the smaller piston. What is the minimum force that must be applied to the smaller piston in order to lift a 3,000-pound car with the larger piston?

(1) 3 pounds
(2) 30 pounds
(3) 100 pounds
(4) 300 pounds
(5) 400 pounds

16. Power brakes in a car operate on the same principle that governs a hydraulic jack. Lightly pushing a brake pedal transmits a force to wheel cylinders that cause the brakes to slow the car.

What would most likely be the result if the hydraulic brake lines were leaking?

(1) The brakes would squeak.
(2) The brakes would be applied all the time.
(3) The brake pedal would lock in place and not move.
(4) Too little pressure would be applied to the brakes.
(5) Excess pressure would be applied to the brakes.

17. Pascal's law helps explain the operation of a hydraulic jack. This law states that a fluid in a container transmits pressure equally in all directions. Which of the following can be given as experimental proof of Pascal's law?

A. When salt is poured into water, a salt-water solution is formed.
B. When water is heated to a temperature of 212°F, the water will boil.
C. When the bottom of an open toothpaste tube is squeezed, toothpaste comes out of the top.

(1) A only
(2) B only
(3) C only
(4) Both A and B
(5) Both B and C

Questions 18 and 19 refer to the following passage, illustration, and graph.

A block slides without friction down a slick ramp. The graph shows the values of both the potential energy (energy at rest) and kinetic energy (energy of motion) of the block as it slides. (*d* is the point along the ground that is directly below the sliding block. The value of *d* slowly changes from 0 to 10.)

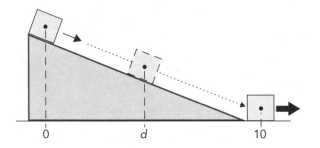

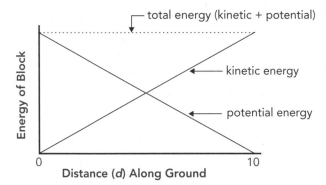

18. At what value of *d* is the kinetic energy of the block about equal to its potential energy?

    (1) *d* = 0
    (2) *d* = 2
    (3) *d* = 5
    (4) *d* = 7
    (5) *d* = 10

19. Which statement about the block is <u>not</u> true?

    (1) The total energy does not change.
    (2) All of the block's potential energy is changed to kinetic energy.
    (3) When *d* = 0, the kinetic energy is 0.
    (4) When *d* = 10, the potential energy is 0.
    (5) All of the block's kinetic energy is changed to potential energy.

Question 20 refers to the information below.

Fact 1: When sunlight is absorbed by an object, the object becomes warmer.
Fact 2: Dark-colored objects absorb more sunlight than light-colored objects.

20. On a hot, sunny day how can you keep coolest?

    (1) by wearing dark-colored clothes and staying out of direct sunlight
    (2) by wearing light-colored clothes and staying in direct sunlight
    (3) by wearing dark-colored clothes and staying in direct sunlight
    (4) by wearing light-colored clothes and staying out of direct sunlight
    (5) by wearing a light-colored jacket over dark-colored clothes and staying out of direct sunlight

21. Which observation is the best evidence that the creation of light is related to the movement of electrons?

    (1) Plants in a cave will carry on photosynthesis in the presence of an electric light.
    (2) Both sunlight and free electrons can travel through the vacuum of space.
    (3) A lightbulb glows when electric current flows through its filament.
    (4) Sunlight is composed of a spectrum of colors.
    (5) The heat created by an electric current can cause a match to light.

22. Luminous objects produce and give off visible light. Objects that only reflect light or that give off light only in response to electric current are not considered luminous. Which of the following is a luminous object?

    (1) a planet
    (2) a spaceship
    (3) the Moon
    (4) a star
    (5) a manmade satellite

Questions 23–26 refer to the passage and graph below.

The graph below refers to an object starting at rest and then falling straight down due to Earth's gravity. The dotted line represents the velocity *(v)* the object reaches when it has fallen a distance *(d)*. This graph refers to small heavy objects for which air resistance does not affect the motion over short distances.

For example, a small rock dropped off a bridge reaches a speed of about 67 feet per second when it has fallen 70 feet.

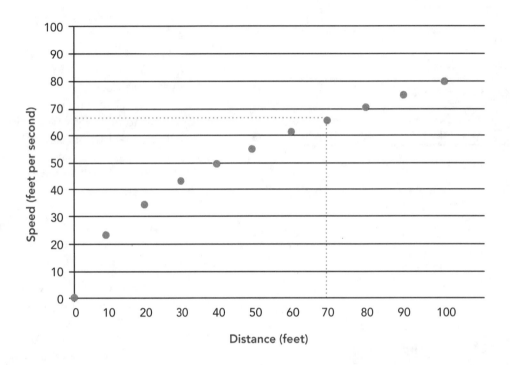

23. Which object would be <u>least</u> likely to fall at a speed indicated on the graph?

    (1) leaf
    (2) brick
    (3) pen
    (4) marble
    (5) bottle

24. A pencil falls from the pocket of a person on a roller coaster. The pencil hits the ground at a speed of 80 feet per second. About how far above the ground was the roller coaster when the pencil fell?

    (1) 50 feet
    (2) 65 feet
    (3) 73 feet
    (4) 81 feet
    (5) 100 feet

25. A stone is dropped off a bridge that is 50 feet above a river. What is the approximate speed of the stone when it hits the water?

    (1) 40 feet per second
    (2) 56 feet per second
    (3) 76 feet per second
    (4) 80 feet per second
    (5) 100 feet per second

26. What limits the eventual speed of an object that is dropped from a very high altitude?

    (1) volume of the object
    (2) weather conditions
    (3) air resistance
    (4) wind speed
    (5) weight of the object

Questions 27–31 refer to the following properties of waves.

Straight-line movement—traveling in a straight line from a source

Reflection—reversing the direction of travel after striking a smooth surface

Refraction—changing the angle of forward motion while moving across a boundary. The refraction is caused by different wavelengths bending by different amounts. When visible light refracts, a spectrum of colors may be seen.

Diffraction—spreading out into a region behind or around a barrier

Wavelength—series of crests and troughs; the distance between two crests (or troughs) is the wavelength.

27. When you hear a child shout from the other side of the house, the sound travels through doorways and around corners in order to reach you. What name is given to this property of sound waves?

    (1) straight-line movement
    (2) reflection
    (3) refraction
    (4) diffraction
    (5) wavelength

28. After a rainstorm you can sometimes see a rainbow of colors displayed in a drop of oil on the street. What name is given to this property of light waves?

    (1) straight-line movement
    (2) reflection
    (3) refraction
    (4) diffraction
    (5) wavelength

29. The sound of a person walking down an empty hallway echoes off the walls of the hall. What property of sound waves is exemplified by echoes?

    (1) straight-line movement
    (2) reflection
    (3) refraction
    (4) diffraction
    (5) wavelength

30. Emilio, a fisherman, spends most of his day sitting in a boat on the ocean. His boat bobs up and down about every seven seconds. What name is given to this property of water waves?

    (1) straight-line movement
    (2) reflection
    (3) refraction
    (4) diffraction
    (5) wavelength

31. What property of sunlight gives rise to the phenomenon that is shown below?

    (1) straight-line movement
    (2) reflection
    (3) refraction
    (4) diffraction
    (5) wavelength

Question 32 is based on the illustration below.

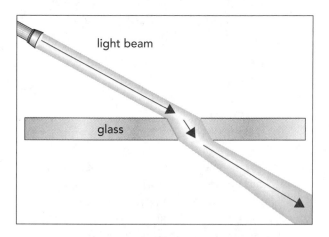

light beam

glass

32. Which of the following facts is the illustration above most likely intended to indicate?

(1) Light changes speed when it passes from one type of material to another.
(2) Light changes direction when it passes from one type of material to another.
(3) The speed of light is greater in glass than in air.
(4) Glass has greater density than air.
(5) Light moves in a straight line through all types of material.

Question 33 refers to the following illustration.

33. What property of nature does the bicycle brake shown above use to advantage?

(1) electricity
(2) resistance
(3) magnetism
(4) energy
(5) gravity

Questions 34 and 35 refer to the following passage.

Socks removed from a hot dryer often cling to each other. Tiny sparks and a crackling noise may occur as you pull them apart. The socks aren't magic, they are just demonstrating the properties of static electricity.

When socks rub together in a dryer, electrons are rubbed off one sock and are transferred to the other. As a result one sock has an excess of electrons and is negatively charged. The other sock has fewer than its normal number of electrons and becomes positively charged—being left with numerous positively charged atoms. The separated negative and positive charges attract one another and the socks cling together. The sparks and crackling sound occur as electrons on the moving socks find the opportunity to jump between socks and join the positively charged atoms.

34. What can you infer from the passage is the best definition of static electricity?

(1) heated electrical charges
(2) separated electrical charges
(3) groups of electrical charges
(4) charged atomic particles
(5) electricity used in appliances

35. What natural phenomenon is most likely a result of static electricity?

(1) solar flares
(2) earthquakes
(3) magnetism
(4) hurricanes
(5) lightning

Question 36 is based on the illustration below.

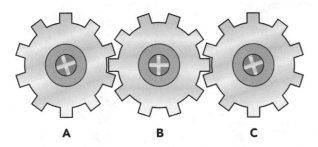

A   B   C

Each gear above can turn clockwise or counterclockwise.

36. Which statement about these gears is true?

 (1) Gears A and B always turn in the same direction.
 (2) If gear B turns clockwise, gear C will also turn clockwise.
 (3) If gear A turns counterclockwise, gear C will also turn counterclockwise.
 (4) If gear A turns clockwise, gear C will turn counterclockwise.
 (5) Gears B and C always turn in the same direction.

Question 37 refers to the following illustration.

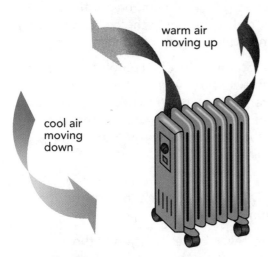

warm air moving up

cool air moving down

37. What name is given to the type of heat flow shown in the illustration?

 (1) conduction
 (2) radiation
 (3) convection
 (4) contraction
 (5) insulation

Questions 38 and 39 refer to the following passage and illustration.

In 1813 the Danish physicist and chemist Hans Christian Ørsted predicted that a connection soon would be found between electricity and magnetism. In 1820 he found that connection himself when he discovered that a current-carrying wire was surrounded by a magnetic field. The strength of the field depended on the distance from the wire and on the amount of electric current.

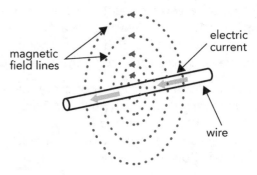

magnetic field lines

electric current

wire

Ørsted's discovery helped scientists understand electric current, which they believed was a type of electric fluid. However, electrons, the particles that produce electric current, wouldn't be discovered until seventy-five years later.

38. What can you infer that scientists today believe gives rise to magnetism?

 (1) the number of copper atoms in a wire
 (2) the strength of a magnetic field
 (3) the direction of electrostatic force
 (4) the movement of electrons
 (5) the number of electrons in an atom

39. Which of the following facts would have been <u>least</u> likely to help Ørsted convince other scientists of his discovery?

 (1) A current-carrying wire does not gain weight as more current flows through it.
 (2) A current-carrying wire causes a nearby compass to deflect.
 (3) An iron nail, surrounded by a current-carrying wire, attracts other iron.
 (4) A current-carrying coil of wire attracts a magnet.
 (5) Two current-carrying coils of wire attract each other.

Questions 40 and 41 refer to the following passage.

An electric generator is a device that converts mechanical energy into electric energy. (An electric generator is also called an *alternator* or a *dynamo*.) Electric motors are devices that convert electric energy into mechanical energy. Two related principles of electromagnetism govern the operation of electric generators and motors.

• The principle of electromagnetic reaction was first observed by French physicist André Ampère in 1820: If an electric current is passed through a wire that is sitting in a magnetic field, the field exerts a force on the wire.

• The principle of electromagnetic induction was discovered by British physicist Michael Faraday in 1831: If a closed loop of wire is moved through a magnetic field, an electric current will flow in the wire.

Many features of modern society, including electricity, radio, television, electric appliances, and computers would not exist without knowledge of the principles of electromagnetism.

40. Which is a result of the principle of electromagnetic reaction?

    (1) The north poles of two magnets repel one another.
    (2) A current-carrying wire twists when placed in a magnetic field.
    (3) An electric current starts to flow in a gold ring as a magnet moves by the ring.
    (4) A current-carrying wire causes the needle of a compass to deflect.
    (5) A current-carrying coil of wire is surrounded by a magnetic field.

41. Which of the following items does <u>not</u> contain an electric generator?

    (1) a microwave oven
    (2) a nuclear power plant
    (3) an automobile
    (4) a power lawnmower
    (5) a hydroelectric plant

Question 42 is based on the following passage and diagram.

A direct-current (DC) motor changes electrical energy into mechanical energy.

ELECTRIC MOTOR

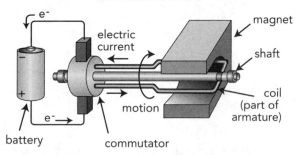

In the cutaway of the motor shown above, an electric current continuously flows in one direction to and from the battery. In the first half of the rotation cycle, current flows in the direction shown by arrows in the coil. The coil is located in a constant magnetic field and, because of the electric current, experiences a force that causes the coil to rotate.

During the second half of the cycle, the electric current flows in the opposite direction in the coil. This ensures that the force continues to rotate the coil in the same direction as during the first half of the cycle. The reversal of current in the coil is caused by the commutator, a type of electrical switch.

The rotating part of a complete motor is called an *armature*. An armature consists of an iron core around which many coils of wire are wrapped. A shaft is connected to the armature so that the rotational energy of the armature can be used.

42. In a DC motor what is the purpose of the commutator?

    (1) to reverse the direction of the magnetic field in which the coil rotates
    (2) to reverse the direction of the current flowing in the coil
    (3) to reverse direction of the coil's rotation
    (4) to decrease the flow of current in the coil
    (5) to increase the flow of current in the coil

Questions 43 and 44 refer to the following illustration.

During flight four forces act on an airplane: lift, gravity, thrust, and drag. For the airplane to stay at the same altitude and fly at a constant speed, the net force on the airplane must be zero.

• The thrust (speed) must be equal in magnitude to the drag, and it must be in the opposite direction.

• The lift must be equal in magnitude to the force of gravity (weight of the airplane), and it must be in the opposite direction.

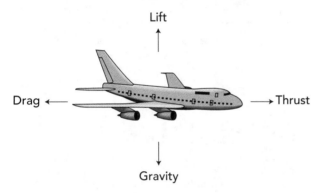

43. What causes drag and thrust forces on an airplane in flight?

  (1) Jet engines cause drag; air resistance causes thrust.
  (2) Air resistance causes drag; jet engines cause thrust.
  (3) Pressure on the bottom of the wings causes thrust; jet engines cause drag.
  (4) Pressure on the bottom of the wings causes drag; gravity causes thrust.
  (5) Air resistance causes drag; lift causes thrust.

44. How does a pilot make an airplane gain altitude (height above ground)?

  (1) Increase the speed of the airplane and reduce the drag force.
  (2) Decrease the weight of the airplane to a value less than the lift force.
  (3) Increase the weight of the airplane to a value greater than the lift force.
  (4) Decrease the lift force to a value less than the weight of the airplane.
  (5) Increase the lift force to a value greater than the weight of the airplane.

Question 45 refers to the passage and illustration below.

For an electric current to flow, there must be a complete electric circuit—a complete path through which the electrons can flow from one side of the power source to the other. In the illustration below, both bulbs will be lit when all three switches are in the closed position.

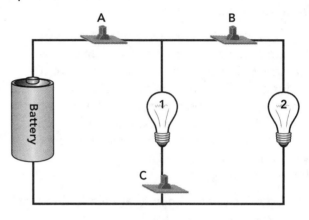

45. With which switch could you open and turn off light 1 but not light 2?

  (1) switch A
  (2) switch B
  (3) switch C
  (4) either switch A or switch C
  (5) either switch B or switch C

Question 46 refers to the following illustration.

46. For what purpose is this tool designed?

  (1) measuring weight
  (2) measuring electric current
  (3) measuring distance
  (4) measuring volume
  (5) measuring temperature

Questions 47 and 48 refer to the following passage.

Scientists in the 1700s knew that when two objects were brought into contact, a process took place that caused the two objects to reach the same temperature. For example, if a hot piece of iron were placed against a cold piece of iron, each piece would soon feel warm. Each piece of iron would feel equally warm (or cool).

To explain this curious fact, scientists at that time concluded that some form of substance flowed from a hotter object to a cooler one. This substance was given the name "caloric." The strange thing was that caloric apparently had no weight. A hot object did not lose weight as caloric flowed from it. A cool object did not gain weight as caloric flowed into it.

Today, scientists know that caloric is in fact heat—energy due to atomic or molecular motion. Scientists also know that heat can transfer from one object to another, be stored, and do work. But heat is not a material substance and does not have weight. Heat is an energetic state of matter, not matter itself.

47. What surprising property did eighteenth-century scientists discover about caloric?

   (1) Caloric is a form of energy.
   (2) Caloric flows from one object to another.
   (3) Caloric has no weight.
   (4) Caloric is contained in food.
   (5) Caloric is a measure of temperature.

48. What word in common use today comes from the word *caloric?*

   (1) the word used to refer to the height of an airplane above the ground
   (2) the word that represents the units of measure on a thermometer
   (3) the word used to indicate the distance around a figure
   (4) the word that represents the rate of energy use of a lightbulb
   (5) the word used as a measure of food energy

Questions 49 and 50 refer to the following illustrations.

In the illustrations below arrows are used to represent the kinetic energy (energy of motion) of individual gas atoms.

A container of low-temperature gas is brought into contact with a container of high-temperature gas.

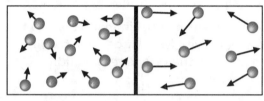

Low Temperature   High Temperature

Later that day

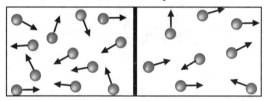

Each container is at the same temperature.

49. What can you infer is actually measured when the temperature of a gas is found?

   (1) the average number of gas atoms or molecules in a container
   (2) the average size of gas atoms or molecules
   (3) the average number of times that gas atoms or molecules strike the wall of a container each second
   (4) the average kinetic energy of gas atoms or molecules
   (5) the average distance that gas atoms or molecules travel before colliding with each other

50. What process brings the two gases shown above to the same temperature?

   (1) the equal distribution of gas particles
   (2) the equal distribution of energy
   (3) the mixing of the gases
   (4) the cooling of the gases
   (5) the absorption of light energy

Answers are on page 113.

GED Science pages 339–368
Complete GED pages 497–532

Directions: Choose the <u>one best answer</u> to each question.

Questions 1–4 refer to the following passage.

Many common minerals can be identified by one or more of the following five properties:

- *Color*—depends on the element(s) that compose a mineral or the impurities in it

- *Luster*—the degree to which a mineral reflects light; a mineral may be classified as metallic or nonmetallic; a mineral with a metallic luster shines like a metal.

- *Hardness*—how easily a mineral can be scratched

- *Streak*—the color of the powder that a mineral leaves when it is rubbed against a hard surface

- *Cleavage*—the way a mineral breaks or splits; cleavage depends on the pattern of the mineral's crystals.

1. Gold prospectors were often fooled by pyrite, or "fool's gold," a mineral that is bright and shiny like gold. However, there is an easy way to identify pyrite—when rubbed against a rock, it leaves a black mark.

   What distinguishing property can be used to identify pyrite?

   (1) color
   (2) luster
   (3) hardness
   (4) streak
   (5) cleavage

2. Silver and copper can be cleaned to the point where you can see your reflection in them. What is this property called?

   (1) color
   (2) luster
   (3) hardness
   (4) streak
   (5) cleavage

3. Corundum, a mineral that is naturally clear, becomes a red ruby color when a small amount of chromium is present. Traces of iron and titanium, though, turn a clear crystal of corundum into a blue sapphire. Which property of corundum is most affected by traces of other substances?

   (1) color
   (2) luster
   (3) hardness
   (4) streak
   (5) cleavage

4. Because of its crystalline structure, a diamond can be split in four separate directions. This allows a diamond to be cut in the shape of a pyramid that brilliantly reflects light in several directions at once. What property determines the shape into which a diamond can be cut?

   (1) color
   (2) luster
   (3) hardness
   (4) streak
   (5) cleavage

Questions 5 and 6 refer to the chart below.

MOHS SCALE OF MINERAL HARDNESS		
Mineral	*Hardness*	*Property*
Talc Gypsum	1 2	can be scratched by a fingernail
Calcite	3	can be scratched by a copper penny
Fluorite Apatite	4 5	can be scratched by a piece of glass
Feldspar Quartz Topaz Corundum	6 7 8 9	can scratch a piece of glass or a knife blade
Diamond	10	can scratch all other common materials

5. According to the Mohs scale, which mineral could <u>not</u> be used in a rubbing compound designed to smooth the edges of a roughly cut piece of glass?

   (1) fluorite
   (2) quartz
   (3) diamond
   (4) feldspar
   (5) topaz

6. According to the Mohs scale, between what two numbers is the hardness rating of a piece of glass?

   (1) 2 and 3
   (2) 3 and 4
   (3) 4 and 5
   (4) 5 and 6
   (5) 6 and 7

Question 7 refers to the following map.

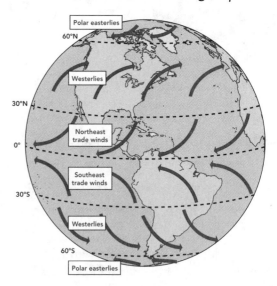

7. Why does a flight from Seattle to New York City take less time than a flight from New York City to Seattle?

   (1) The time in New York City is three hours earlier than the time in Seattle.
   (2) The time in New York City is three hours later than the time in Seattle.
   (3) The flight path from Seattle to New York City is much shorter than the return path.
   (4) The prevailing winds increase the airplane speed relative to the ground speed on the flight from Seattle to New York City.
   (5) The prevailing winds increase the airplane speed relative to the ground speed on the flight from New York City to Seattle.

8. Which of the following is <u>least</u> related to the fact that Earth is spherical in shape?

   (1) A ship disappears from sight as it sails away from shore.
   (2) The position of a star is different when viewed from a northern state than when viewed from a southern state.
   (3) During a lunar eclipse Earth's shadow appears on the Moon as a curved line.
   (4) Both the Sun and a full moon have circular outlines when viewed in the sky.
   (5) People in China see the Sun come up at the same time that people in the United States see the Sun go down.

Questions 9–11 refer to the illustration and passage below.

OCEAN FLOOR

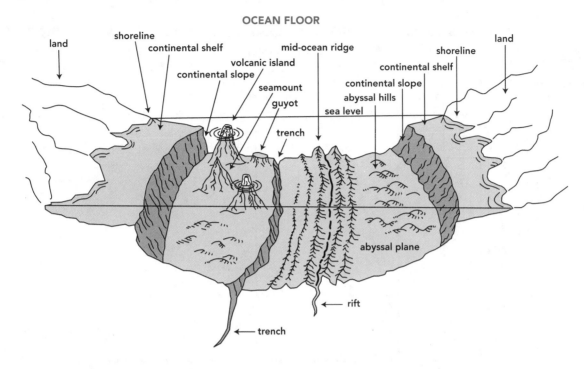

The Mid-Atlantic Ridge is a long underwater mountain range that sits about halfway between the continents on either side of it.

Scientists believe that the sea floor is spreading outward along this ridge. The spreading seems to be caused by the continual flow of magma from cracks in Earth's crust and from eruptions in volcanoes along this mountain chain. Because of the sea floor spreading, North and South America are slowly moving farther from Europe and Africa.

9. What is the probable cause of the spreading of the sea floor at the Mid-Atlantic Ridge?

   (1) the movement of continents, followed by underwater earthquakes
   (2) the movement of ocean currents due to the rotation of Earth on its axis
   (3) the formation of mountain ranges due to faults in underwater land formations
   (4) the movement of tectonic plates with an accompanying outpouring of magma
   (5) the eruption of underwater volcanoes and an outpouring of lava

10. What is the name of the deepest part of the ocean?

   (1) seamount
   (2) rift
   (3) continental slope
   (4) abyssal plain
   (5) trench

11. What is the best evidence that the sea floor spreading occurs at about the same rate on both sides of the ridge?

   (1) The shape of South America is similar to the shape of Africa.
   (2) The continents to the left of the ridge are about the same distance from the ridge as the continents to the right.
   (3) The sea floor to the left of the ridge contains the same types of minerals and food as the sea floor to the right.
   (4) The ridge is the same general shape as the shorelines on each side of it.
   (5) Many types of minerals found in South America are also found in Africa.

Questions 12–14 are based on the following graph.

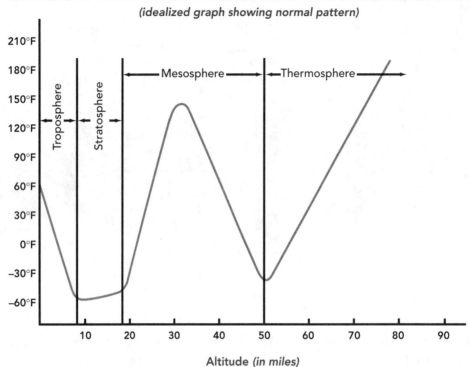

CHANGES OF ATMOSPHERIC TEMPERATURE WITH INCREASING ALTITUDE
(idealized graph showing normal pattern)

12. Between what two layers of atmosphere does the lowest temperature occur?

(1) mesosphere and thermosphere
(2) troposphere and mesosphere
(3) troposphere and stratosphere
(4) stratosphere and mesosphere
(5) stratosphere and thermosphere

13. In which layer(s) of atmosphere does the air temperature at first rise with increasing altitude and, at higher altitudes, decrease as the altitude increases?

(1) troposphere
(2) stratosphere
(3) mesosphere
(4) thermosphere
(5) troposphere and thermosphere

14. A commercial airplane traveling cross-country cruises at an altitude of 33,000 feet (about six miles). How does the air temperature just outside the plane change between the time the plane takes off and the time it reaches its cruising altitude?

(1) The temperature continually increases.
(2) The temperature continually decreases.
(3) The temperature at first decreases and then increases.
(4) The temperature at first increases and then decreases.
(5) The temperature remains approximately constant.

Questions 15–17 refer to the following passage.

Did you ever wonder why cloudy nights are warmer than clear, starlit nights? The answer has to do with how the surface of Earth is heated and cooled.

During the day, Earth's surface absorbs much of the sunlight that strikes it. The absorbed light heats land and water masses in much the same way that sunlight warms you. The hotter the day, the more sunlight energy is absorbed, and the hotter the surface becomes.

When the Sun goes below the horizon, the air quickly cools. Land and water, however, cool more slowly because they hold much more heat energy. This heat energy slowly radiates upward, away from Earth's surface.

- On a clear night most of this energy radiates back into space and is lost.

- On a cloudy night clouds absorb heat energy that radiates upward from the surface. The result is that a layer of clouds acts like a blanket and traps heat energy between the clouds and Earth's surface, warming the air in between.

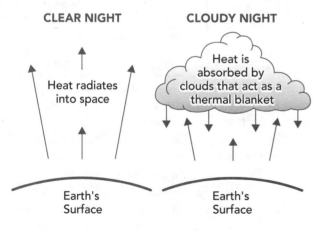

15. What is the key point made in the passage?

(1) Clouds trap heat that radiates from Earth's surface at night.
(2) Earth is warmed by sunshine that strikes its surface.
(3) In the evening air cools more quickly than land or water.
(4) Clouds form only on warm evenings.
(5) Clouds trap heat that radiates from the Sun to Earth.

16. In which of the following places, all having the same average daytime temperature, would you expect the average nighttime temperature to be the lowest?

(1) along the ocean shore
(2) in a tropical rain forest
(3) next to one of the Great Lakes
(4) in a desert
(5) on a small island

17. For which of the following conditions would the nighttime temperatures most likely be the highest?

(1) a totally cloudy day, a totally cloudy night
(2) a partly cloudy day, a totally cloudy night
(3) a clear day, a totally cloudy night
(4) a clear day, a partly cloudy night
(5) a partly cloudy day, a totally clear night

Questions 18–20 refer to the passage and illustration below.

Soil drainage depends on the relative amounts of clay, silt, and sand contained in the soil. Clay particles are typically a few thousandths of a millimeter wide; silt particles are a few hundredths of a millimeter wide; and sand particles are a few tenths of a millimeter wide.

A soil scientist is conducting an experiment to test the drainage of different soils obtained from a riverbed. The scientist places soil samples in each of six containers. He then measures the drainage properties of the soil samples by pouring an equal amount of water in each container. Next, he measures the rate at which water flows out the screened bottom of each container.

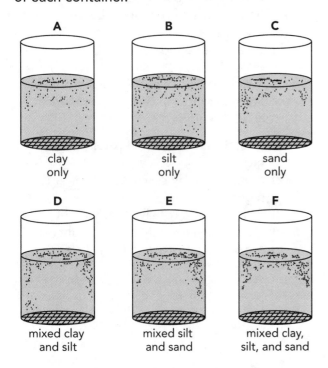

18. To compare the drainage property of silt with that of sand, which two containers should the scientist test?

    (1) A and C
    (2) B and C
    (3) B and F
    (4) C and D
    (5) D and E

19. Which of the following conditions must be met if the experiments are to give meaningful results?

    A. Each container must contain an equal weight of soil.
    B. Before water is added, the soil in each container must be dry.
    C. The depth of the soil must be the same in each container.

    (1) A only
    (2) B only
    (3) Both A and C
    (4) Both B and C
    (5) A, B, and C

20. The results of the experiment with containers A, B, and C are shown on the graph below. Which of the following conclusions is best supported by information given in the passage and on the graph?

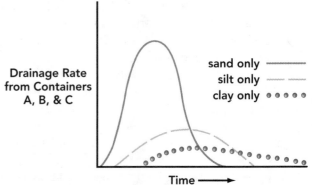

    (1) The smaller the soil particles, the better the drainage.
    (2) The larger the soil particles, the better the drainage.
    (3) Drainage is better when soil particles are tightly packed together.
    (4) Clay drains more quickly than silt does.
    (5) Not enough information is given to draw any conclusions about drainage.

Questions 21–24 refer to the following passage and illustration.

The discovery that Earth acts as a giant magnet led to the development of the magnetic compass, a useful direction-finding device. The most popular compass is a case containing a magnetized needle that rotates around its center point. The needle is a narrow bar magnet and has both a north pole and a south pole. You simply hold this compass in your hand and watch the needle line up with the direction of Earth's magnetic field. From the direction the needle points, you can determine the direction you're traveling. You can also use the compass to determine what direction one location is from a second location.

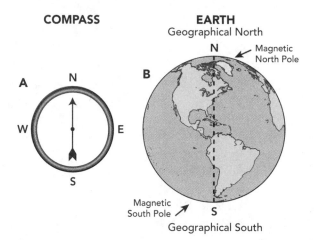

COMPASS

EARTH
Geographical North

Magnetic North Pole

Magnetic South Pole

Geographical South

Scientists have discovered that the north pole of a compass needle always points toward Earth's north magnetic pole. The north magnetic pole—so named because it is located in the Northern Hemisphere—is a point about 1,200 miles from the geographical North Pole (the northernmost point on Earth).

Similarly, the south pole of a compass needle always points toward Earth's south magnetic pole, a point located in Antarctica, about 1,200 miles from the geographical South Pole (the southernmost point on Earth).

Because scientists know that magnetic fields can be produced by electricity, they believe that Earth's magnetic field is caused by electric currents within Earth, possibly related to the flow of liquid iron and other metals in the liquid part of Earth's interior.

21. According to the passage, what do scientists believe causes Earth's magnetic field?

   (1) deposits of magnetic rock
   (2) the development of the compass
   (3) the force of gravity
   (4) electric currents within Earth
   (5) the two magnetic poles

22. Which of the following people would have the <u>least</u> use for a compass?

   (1) a mountain climber
   (2) a mechanic
   (3) an explorer
   (4) a mapmaker
   (5) a pilot

23. When is a hiker's compass most likely to give an incorrect reading?

   (1) when the hiker is in a strong wind
   (2) when the hiker reaches the top of a high mountain
   (3) when the hiker walks along the shore of a large lake
   (4) when the hiker is in a rainstorm
   (5) when the hiker passes beneath electric power lines

24. Which of the following statements can you infer to be true?

   (1) Earth's north magnetic pole is the south pole of Earth's magnetic field.
   (2) Earth's north magnetic pole is the north pole of Earth's magnetic field.
   (3) Earth's magnetic field does not have a north pole or a south pole.
   (4) Earth's magnetic field has two north poles and two south poles.
   (5) Earth's magnetic field is strongest at the equator.

Questions 25–27 refer to the following passage and graph.

Earth periodically experiences an ice age, a time of extreme cold when glaciers cover most of the land. As Earth's temperature falls, glaciers get larger and ocean levels drop. Evaporating ocean water blows over the land and falls to Earth as freezing rain and snow. Since the atmosphere is so cold, little of the ice melts to refill the oceans. During an ice age, the sea level drops. At the end of an ice age, the sea level gradually returns to normal level.

Recently, scientists have found evidence indicating that the average sea level has varied over the past 35,000 years as shown on the graph below.

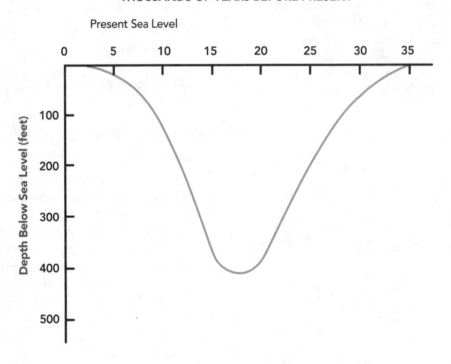

THOUSANDS OF YEARS BEFORE PRESENT

25. At about what time period shown on the graph did the sea level reach its lowest point?

(1) within the last 1,000 years
(2) about 10,000 years ago
(3) about 18,000 years ago
(4) about 26,000 years ago
(5) more than 35,000 years ago

26. How many ice ages have most likely occurred on Earth in the last 35,000 years?

(1) 0
(2) 1
(3) 2
(4) 3
(5) 4

27. Which conclusion is best supported by this graph?

(1) The most recent ice age ended about 18,000 years ago.
(2) The most recent ice age began about 18,000 years ago.
(3) The sea level has stayed nearly constant during the past 35,000 years.
(4) Scientists have been monitoring the sea level for 35,000 years.
(5) During the past 18,000 years the sea level has been gradually dropping.

Questions 28 and 29 refer to the graph below.

The graph shows four temperature readings taken on Sierra Peak, a 10,540-foot-high mountain.

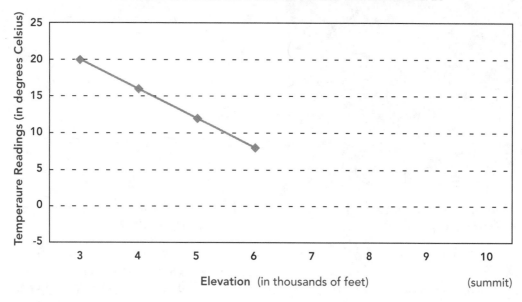

NOVEMBER 9 TEMPERATURE READINGS ON SIERRA PEAK

28. At about what rate is the temperature dropping as altitude increases?

(1) 1°C per 1,000 feet
(2) 2°C per 1,000 feet
(3) 3°C per 1,000 feet
(4) 4°C per 1,000 feet
(5) 5°C per 1,000 feet

29. Assuming a constant rate of temperature decrease with increasing altitude, at what altitude will the freezing level (0°C) occur?

(1) below 6,000 feet
(2) about 6,000 feet
(3) about 8,000 feet
(4) about 10,000 feet
(5) above 10,000 feet

30. What process is occurring when tons of rocks slide from the top of a cliff to the bottom?

(1) physical weathering
(2) chemical weathering
(3) wind erosion
(4) water erosion
(5) gravity erosion

31. Which of the following is <u>not</u> characteristic of both the eruption of a geyser and the eruption of a volcano?

(1) a heat source of hot magma
(2) a flow of liquid through Earth's surface
(3) a cone that forms near the top of a flow tube
(4) a high-pressure region of hot gases
(5) a fairly predictable eruption schedule

Question 32 refers to the following photograph.

Mushroom-shaped rocks, such as the one pictured below, are often seen in deserts.

32. What is the most likely cause of the shape of mushroom-shaped rocks?

   (1) erosion from the impact of wind-blown sand
   (2) erosion from the action of rainwater flowing through the sand
   (3) lightning strikes hitting the sand near the base of the rocks
   (4) magma flowing through the sand
   (5) erosion from the movement of nearby sand due to periodic earthquakes

Question 33 refers to the following illustration.

33. What is the anemometer, pictured above, designed to measure?

   (1) earthquake strength
   (2) water speed
   (3) wind speed
   (4) rainfall amount
   (5) temperature

Questions 34 and 35 refer to the following graph.

MAXIMUM WATER VAPOR CONTENT OF AIR

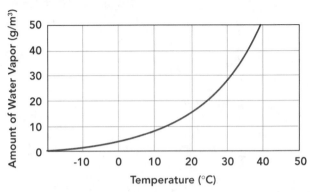

34. What general conclusion can you draw from the graph?

   (1) The maximum moisture content of air does not depend on air temperature.
   (2) The temperature of air increases as its moisture content increases.
   (3) The temperature of air decreases as its moisture content increases.
   (4) The maximum moisture content of air decreases as air temperature increases.
   (5) The maximum moisture content of air increases as air temperature increases.

35. Relative humidity is the amount of moisture that air holds, expressed as a percent of the maximum amount it is able to hold.

Suppose the air temperature in your city is 20°C and the relative humidity is 25 percent. About how much water vapor is in the air?

   (1) 1 gram per cubic meter
   (2) 4 grams per cubic meter
   (3) 12 grams per cubic meter
   (4) 15 grams per cubic meter
   (5) 60 grams per cubic meter

Answers are on page 115.

Space Science

GED Science pages 369–384
Complete GED pages 497–532

Directions: Choose the <u>one best answer</u> to each question.

1. A meteor, often called a shooting star, is a brief streak of light across the night sky. A meteor is not really a star at all. It is a small piece of fast-moving matter from space, a meteoroid, that burns up upon entering Earth's upper atmosphere.

   Which statement below is an opinion rather than a scientific fact?

   (1) Many meteors can be seen when Earth passes a point in space where a comet has passed.
   (2) Shooting stars bring luck to those who see them.
   (3) In some cultures meteors are a religious symbol.
   (4) Many pieces of matter in space are tiny fragments of ice and dust left behind by passing comets.
   (5) A few meteors pass through the atmosphere and strike Earth's surface.

2. The constellations (patterns of stars) seen in the Northern Hemisphere are not the same those seen in the Southern Hemisphere.

   Which statement below best explains this observation?

   (1) Countries in the Northern Hemisphere experience summer at the same time when countries in the Southern Hemisphere experience winter.
   (2) Constellations have been studied since ancient times.
   (3) The northern half of Earth points toward a different part of space than the southern half of Earth.
   (4) The noonday sun is higher in the sky in summer than in winter.
   (5) The distance between Earth and the Sun is not the same in winter as it is in summer.

Questions 3 and 4 refer to the passage and illustration below.

As shown in the illustration below, Polaris (called the North Star) is almost directly in line with Earth's rotation axis.

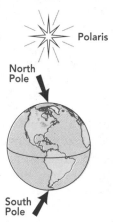

3. At which of the following locations would the North Star appear to be in the sky directly above a person standing on Earth?

   (1) the North Pole
   (2) halfway between the equator and the North Pole
   (3) the equator
   (4) halfway between the equator and the South Pole
   (5) the South Pole

4. To use the North Star as a direction-finder, what must you assume to be true?

   (1) The North Star is the brightest evening star.
   (2) The North Star can be located by first locating the group of stars known as the Big Dipper.
   (3) The North Star is as easily seen during daytime as during nighttime.
   (4) The North Star stays in the same position in the sky at all hours of the night.
   (5) The North Star is in the Milky Way galaxy along with the Sun.

Questions 5 and 6 refer to the passage below.

Communications satellites are placed in synchronous orbits around Earth. A satellite in a synchronous orbit stays directly above a particular point on Earth's surface at all times.

5. For a satellite to be in synchronous orbit, how often must it make a complete trip around Earth?

   (1) twice every 24 hours
   (2) once every 24 hours
   (3) twice a month
   (4) once a month
   (5) once a year

6. For which of the following activities would a synchronous-orbit satellite be useful?

   A. Photographing each continent on Earth
   B. Continually transmitting television programs from one country to another
   C. Measuring sun energy while staying in a position directly between Earth and the Sun

   (1) A only
   (2) B only
   (3) C only
   (4) Both A and B
   (5) Both A and C

7. Which is the only type of energy that will not be a potential source of energy for astronauts traveling to Mars in the future?

   (1) chemical energy
   (2) hydroelectric energy
   (3) gravitational energy
   (4) nuclear energy
   (5) solar energy

Questions 8 and 9 refer to the passage and photograph below.

The Hubble Space Telescope (HST) was launched into orbit on April 25, 1990. This telescope is the first orbiting observatory ever placed into space—the region above Earth's atmosphere. The primary mirror of the telescope has a diameter of almost 8 feet.

The HST has enabled scientists to see for the first time many things in the universe. Among these are the collisions of distant galaxies and the discovery of a planet outside of our own solar system.

8. What is the most likely purpose of the flat panels on each side of the cylindrical HST?

   (1) to produce electricity from moonlight
   (2) to produce electricity from sunlight
   (3) to photograph Earth's surface
   (4) to serve as a beacon for astronauts
   (5) to protect the HST from space debris

9. What main advantage to scientists does the HST have over ground-based telescopes?

   (1) ability to point to any part of the universe
   (2) reduced maintenance requirements
   (3) solar power
   (4) position above Earth's atmosphere
   (5) easy access by space shuttle

Questions 10–12 refer to the following passage and photograph.

The Moon is the name given to the only natural satellite of Earth. The diameter of the Moon is about 2,160 miles, about one-third that of Earth, and the average distance of the Moon from Earth is 238,857 miles.

The face of the Moon seen from Earth shows two bright craters: Tycho near the bottom left, and Copernicus directly north of Tycho. The dark areas are knows as seas. The sea at the far left is Oceanus Procellarum; the sea at the top center is Mare Imbrium; the sea at the far right is Mare Crisium; the sea to the left of Mare Crisium is Mare Tranquillitatis.

The first landing of humans on the Moon took place July 20, 1969, when Neil Armstrong and Edwin Aldrin, Jr., landed in Mare Tranquillitatis.

10. Because the Moon has no atmosphere, which of the following does the Moon <u>not</u> experience?

    (1) sunrise and sunset
    (2) solar heating
    (3) meteorite impacts
    (4) wind erosion
    (5) surface temperature changes

11. If Earth had the same diameter as the Moon, a 200-pound man on today's Earth would weigh 50 pounds on this smaller Earth. This same man would weigh only 33 pounds on the Moon. What is the most likely reason this is so?

    (1) The Moon has a hollow center.
    (2) The average density of Earth is less than the average density of the Moon.
    (3) The average density of the Moon is less than the average density of Earth.
    (4) Unlike the Moon, Earth has oceans.
    (5) Unlike the Moon, Earth has an atmosphere.

12. A very interesting fact is that the same side of the Moon always faces Earth. A person standing on Earth never sees the backside of the Moon.

    Which of the following must be true for the same side of the Moon to always face Earth?

    (1) The Moon rotates once on its rotation axis in the same length of time it makes one complete path around Earth.
    (2) The Moon does not rotate as it makes one complete path around Earth.
    (3) The Moon rotates twice on its rotation axis in the same length of time it makes one complete path around Earth.
    (4) Only one side of the Moon is ever lit by sunlight.
    (5) The dark side of the Moon always faces the dark side of Earth.

Questions 13 and 14 refer to the passage below.

One of the strangest objects in the universe is known as a *black hole*. A black hole is believed to be the very dense remains of a large star that, due to gravitational force, has collapsed to a fraction of its previous size. The gravitational field of a black hole is so strong that even light cannot escape from it.

A black hole is surrounded by a spherical boundary, called a horizon. Light can enter the horizon but can never escape. Therefore a black hole appears totally black, its outer radius extending to its horizon.

13. Which of the following is <u>not</u> mentioned as a property of a black hole?

    (1) extremely high density
    (2) exceptionally strong gravitational field
    (3) remains of a collapsed star
    (4) ability to capture light
    (5) tendency to become a supernova

14. How would a black hole appear to an astronomer who tries to view it?

    (1) as a large, bright object in the sky
    (2) as a small, bright object in the sky
    (3) as a dark void in the sky
    (4) as an object blinking on and off
    (5) as an object slowly becoming darker

Questions 15 and 16 refer to the passage and photographs below.

A *solar flare* is a sudden explosion of visible light, particles, and radio-frequency radiation from the Sun's surface. These flares are believed to be caused by the interactions of the Sun's magnetic field and matter on the Sun's surface.

When a solar flare occurs, the increased sunlight is not noticed on Earth because the change in light energy is small compared to the total light given off by the Sun. Scientists get their best view of solar flares during a solar eclipse.

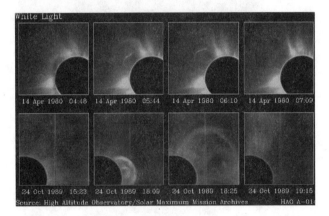

15. What most likely gets disrupted on Earth during intense solar flare activity?

    (1) climate patterns
    (2) radio communications
    (3) flow of ocean tides
    (4) earthquake frequency
    (5) length of daylight hours

16. What makes it possible for scientists to view solar flares during a solar eclipse?

    (1) A solar eclipse causes solar flares.
    (2) The Sun blocks the main part of the Moon from view.
    (3) The Moon blocks the main part of the Sun from view.
    (4) The Moon's gravity is less than Earth's gravity.
    (5) The Moon is smaller than Earth.

Answers are on page 116.

Science

This practice test will give you an opportunity to evaluate your readiness for the GED Science Test.

Directions: Choose the <u>one best answer</u> to each question. The questions are based on reading passages, charts, graphs, maps, and cartoons. Answer each question as carefully as possible. If a question seems to be too difficult, do not spend too much time on it. Work ahead and come back to it later when you can think it through carefully. You should take approximately 80 minutes to complete the test.

Practice Test Answer Grid

1 ① ② ③ ④ ⑤	18 ① ② ③ ④ ⑤	35 ① ② ③ ④ ⑤
2 ① ② ③ ④ ⑤	19 ① ② ③ ④ ⑤	36 ① ② ③ ④ ⑤
3 ① ② ③ ④ ⑤	20 ① ② ③ ④ ⑤	37 ① ② ③ ④ ⑤
4 ① ② ③ ④ ⑤	21 ① ② ③ ④ ⑤	38 ① ② ③ ④ ⑤
5 ① ② ③ ④ ⑤	22 ① ② ③ ④ ⑤	39 ① ② ③ ④ ⑤
6 ① ② ③ ④ ⑤	23 ① ② ③ ④ ⑤	40 ① ② ③ ④ ⑤
7 ① ② ③ ④ ⑤	24 ① ② ③ ④ ⑤	41 ① ② ③ ④ ⑤
8 ① ② ③ ④ ⑤	25 ① ② ③ ④ ⑤	42 ① ② ③ ④ ⑤
9 ① ② ③ ④ ⑤	26 ① ② ③ ④ ⑤	43 ① ② ③ ④ ⑤
10 ① ② ③ ④ ⑤	27 ① ② ③ ④ ⑤	44 ① ② ③ ④ ⑤
11 ① ② ③ ④ ⑤	28 ① ② ③ ④ ⑤	45 ① ② ③ ④ ⑤
12 ① ② ③ ④ ⑤	29 ① ② ③ ④ ⑤	46 ① ② ③ ④ ⑤
13 ① ② ③ ④ ⑤	30 ① ② ③ ④ ⑤	47 ① ② ③ ④ ⑤
14 ① ② ③ ④ ⑤	31 ① ② ③ ④ ⑤	48 ① ② ③ ④ ⑤
15 ① ② ③ ④ ⑤	32 ① ② ③ ④ ⑤	49 ① ② ③ ④ ⑤
16 ① ② ③ ④ ⑤	33 ① ② ③ ④ ⑤	50 ① ② ③ ④ ⑤
17 ① ② ③ ④ ⑤	34 ① ② ③ ④ ⑤	

PRACTICE TEST

Question 1 refers to the illustration below.

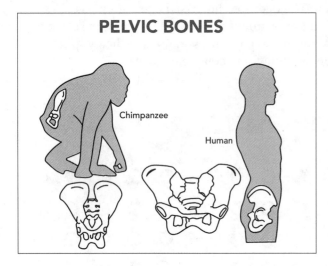

PELVIC BONES

Chimpanzee

Human

1. The shape of the pelvis, or hip bones, in primates is related to the strength and function of the upper leg muscles. Shown above are scale drawings of the pelvis of a human being and a chimpanzee. To what activity is the shape of the pelvis most likely related?

   (1) swinging by use of arms
   (2) eating
   (3) walking upright
   (4) sleeping
   (5) rolling over

2. An object that is *biodegradable* is naturally decomposed (broken down into simpler substances) by living things in the environment. Which of the following would most likely be classified as biodegradable?

   (1) a plastic milk container
   (2) a rusting piece of iron
   (3) a crumbling ancient rock wall
   (4) a fallen dead tree
   (5) an abandoned automobile tire

Question 3 refers to the following passage and illustration.

A magnetic field forms around a wire that is carrying an electric current. The strength of the field is indicated by two features of the circular field lines:

- The number of field lines

- The closeness of the field lines to the wire and to one another

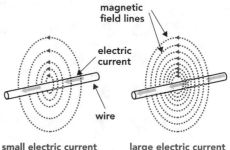

magnetic field lines

electric current

wire

small electric current   large electric current

3. What does the strength of the magnetic field surrounding a current-carrying wire depend on?

   A. the amount of electric current
   B. the length of the wire
   C. the distance of the field lines from the wire

   (1) A only
   (2) B only
   (3) C only
   (4) Both A and B
   (5) Both A and C

4. What is the most likely safety-related purpose of the reflective screen on a microwave's glass door?

   (1) to prevent anyone from seeing into the oven
   (2) to keep microwaves from leaving the oven
   (3) to protect the door against breakage
   (4) to keep heat from leaving the oven
   (5) to keep room light from getting into the oven

PRACTICE TEST

Question 5 refers to the illustration below.

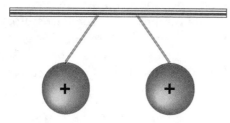

5. Each of two positively charged spheres is hanging by a string. The spheres have two forces acting on them:

   • Gravitational force pulling them together

   • Electrostatic force pushing them apart

   Which statement about the spheres is true?

   (1) Gravitational force is always attractive, and electrostatic force is always repulsive.
   (2) The electrostatic force is not as strong as the gravitational force.
   (3) The electrostatic force is stronger than the gravitational force.
   (4) The electrostatic force is equal in strength to the gravitational force.
   (5) If both spheres were negatively charged, they would be pulled together instead of pushed apart.

6. A study is going to be conducted to try to determine whether mild hypertension is affected by salt in the diet. Hypertension is commonly referred to as high-blood pressure.

   Two groups of people will be studied. The first group will be put on a low-salt, 1,500-calorie-per-day diet. What type of diet should the second group maintain?

   (1) low-salt, 1,000-calorie diet
   (2) low-salt, 1,500-calorie diet
   (3) low-salt, 3,000-calorie diet
   (4) high-salt, 1,000-calorie diet
   (5) high-salt, 1,500-calorie diet

7. Which of the following comments is most likely an opinion that cannot be checked using scientific evidence?

   (1) A country's decision to allow whale hunting is not the concern of any other country.
   (2) Dolphins and porpoises are actually two different types of toothed whales.
   (3) Whale hunting, if not controlled, could lead to the extinction of whales during this new century.
   (4) Whales communicate with one another by a series of cries.
   (5) The largest animal on Earth today is the blue whale, weighing as much as 130 tons.

Question 8 refers to the following diagram.

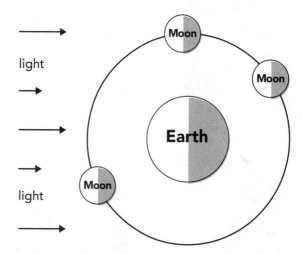

8. What is this model most likely designed to show?

   (1) the early history of Earth
   (2) the movement of Earth around the Sun
   (3) the seasons on the Moon
   (4) the seasons on Earth
   (5) the phases of the Moon

PRACTICE TEST

9. Carbon dioxide gas does not burn. For this reason, it is effective in putting out fires. Which additional fact(s) would enable you to conclude that a fire extinguisher containing carbon dioxide is an excellent household safety device?

   A. Water can cause an oil fire to spread and can increase the toxic fumes of an electrical fire.
   B. Carbon dioxide gas dissolves in water, forming a small amount of carbonic acid, which is useful as a preservative.
   C. Carbon dioxide gas is not harmful to humans.
   D. Pressurized carbon dioxide cools when it flows through the opening of its container and will quickly reduce the temperature of anything it touches.

   (1) A and C
   (2) B and D
   (3) C and D
   (4) A, B, and C
   (5) A, C, and D

Question 10 refers to the following illustration.

INCLINED PLANE

10. Which phrase best describes the principle involved in using an inclined plane?

    (1) more force, shorter distance
    (2) more force, longer distance
    (3) less force, shorter distance
    (4) less force, longer distance
    (5) equal force, different direction

11. An *isomer* is a molecule that has the same number and kinds of atoms as another molecule but has a different structure and different chemical properties. For example, pentane (C_5H_{12}) has several isomers. Shown below is a molecule of the most common isomer called *normal pentane* or *n-pentane*.

$$H - C - C - C - C - C - H$$

Which molecule is an isomer of *n-pentane*?

(1)
$$H - C - C - C - C - H$$

(2)
$$H - C - C - C - C - H$$

(3)
$$H - C - C - C - C - H$$
$$H - C - H$$

(4)
$$H - C - C - C - C - H$$
$$C - H - C$$
$$C$$

(5)
$$H - C - C - C - C - H$$
$$H - C - C - H$$

PRACTICE TEST

Questions 12 and 13 refer to the following passage.

The amount of gas that will dissolve in a liquid depends on two things: the pressure of that gas above the liquid surface and the temperature of the liquid.

Increasing the pressure increases the amount of dissolved gas that a liquid will hold. Decreasing the pressure decreases the amount of dissolved gas that a liquid will hold.

Increasing the temperature of the liquid, however, has the opposite effect: It decreases the amount of dissolved gas a liquid will hold. Decreasing the temperature increases the amount of dissolved gas that a liquid will hold.

12. When a bottle of a carbonated beverage is opened, it *effervesces* (bubbles) as carbon dioxide gas rapidly escapes from the bottle and the liquid.

    When opening a bottle of soda, what change occurs that results in effervescence?

    (1) a rapid decrease in pressure of carbon dioxide gas over the surface of the soda
    (2) a slow increase in the soda's temperature
    (3) a rapid decrease in the soda's temperature
    (4) a rapid increase in pressure of carbon dioxide gas over the surface of the soda
    (5) a rapid increase in the soda's temperature

13. Caps sometimes blow off the tops of ginger-ale bottles when they are exposed to direct sunlight for a long time. To what is this effect most likely related?

    (1) chemical changes in the sugar
    (2) weakening of the bottle caps
    (3) increasing temperature of the ginger ale
    (4) decreasing pressure in the ginger-ale bottles
    (5) expansion of the heated ginger-ale bottles

Questions 14 and 15 refer to the following passage.

Many plants are adapted by nature so that their seeds are dispersed over a great distance. For some plants the seeds are scattered when birds and other animals eat the fruit. Because many types of seeds are protected by a coating that is not easily digested, swallowed seeds pass through an animal's digestive system unharmed. The seeds can be dispersed anywhere the animal travels.

Other types of plants produce fruit that is covered with sharp barbs that make the fruit difficult, if not impossible, to eat. In these plants the barbs attach themselves to an animal's fur. The animal disperses the seeds as it rubs against the ground or a tree in an effort to get rid of the unwanted hitchhikers.

14. What is the most important way that an apple contributes to the growth of new apple trees?

    (1) An apple catches on the fur of passing animals.
    (2) An apple attracts insects to pollinate the tree.
    (3) An apple provides nutritious food for wild animals.
    (4) An apple provides nutrients to the tree after the apple falls.
    (5) An apple attracts animals that eat apples and swallow apple seeds.

15. Which of the following factors is most related to the ability of an animal to disperse the seeds of a mulberry tree over a great distance?

    (1) the speed of the animal while running
    (2) whether the animal flies
    (3) the types of predators that hunt the animal
    (4) whether the animal migrates in winter
    (5) the size of the animal

PRACTICE TEST

Questions 16–18 refer to the graph below.

BREAKDOWN OF EARTH'S LAND

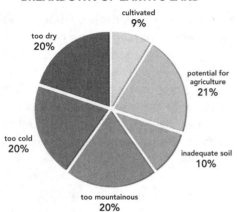

cultivated
9%

too dry
20%

potential for
agriculture
21%

too cold
20%

inadequate soil
10%

too mountainous
20%

16. According to the graph what total percent of Earth's land is either presently cultivated or potentially usable for agriculture?

    (1) 9 percent
    (2) 23 percent
    (3) 30 percent
    (4) 50 percent
    (5) 64 percent

17. In which of the following categories would the state of Iowa be placed?

    (1) actually cultivated
    (2) potentially usable for agriculture
    (3) too mountainous
    (4) too cold
    (5) too dry

18. Which statement is best supported by data shown on the graph?

    (1) Deserts cover more of Earth's surface than do oceans.
    (2) Deserts cover less of Earth's surface than does cultivated land.
    (3) Deserts cover more of Earth's surface than does land that is too cold for cultivation.
    (4) Deserts cover about an equal amount of Earth's surface as does cultivated land.
    (5) Deserts cover an equal amount of Earth's surface as do mountains.

19. Suppose that on Monday you flip a penny and it lands heads up. You then flip the same penny on Tuesday, and it again lands heads up. On Wednesday you flip the penny once more. Which of the following is true about the flip on Wednesday?

    (1) The penny is equally likely to land heads up or tails up.
    (2) The penny is more likely to land heads up than tails up.
    (3) The penny is more likely to land tails up than heads up.
    (4) The penny will land heads up for sure.
    (5) The penny will land tails up for sure.

20. What name is given to the basic unit of all living things?

    (1) energy
    (2) environment
    (3) organ
    (4) response
    (5) cell

21. The ability of a living thing to maintain a stable set of conditions within its body is called *homeostasis*. One example of homeostasis is the ability of warm-blooded animals to maintain a relatively constant body temperature. When a warm-blooded animal becomes too hot, it sweats. Sweating helps an animal to cool and to maintain its normal temperature. Sweating is called a *homeostatic reflex*.

    Which of the following would <u>not</u> be classified as a homeostatic reflex?

    (1) an increase in heart rate during exercise
    (2) shivering after jumping into a cold lake
    (3) gasping for air while running fast
    (4) jumping at the sound of an unexpected loud noise
    (5) feeling hungry after awakening in the morning

PRACTICE TEST

Questions 22 and 23 refer to the following diagrams.

HOW PARENTS PASS ON SICKLE-CELL ANEMIA

Diagram A

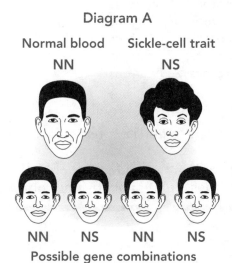

Normal blood — NN
Sickle-cell trait — NS

NN   NS   NN   NS

Possible gene combinations
in a child

NS = Sickle-cell trait
NN = Normal blood
SS = Sickle-cell disease

Diagram B

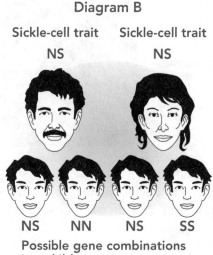

Sickle-cell trait — NS
Sickle-cell trait — NS

NS   NN   NS   SS

Possible gene combinations
in a child

22. Diagram A shows the gene combinations of a child who could be born to the parents illustrated: The father has normal blood (*NN*), and the mother has one sickle-cell gene (*NS*). What is the chance that a child of this couple will also have the sickle-cell trait (an *NS*-gene combination)?

(1) 0 percent (no chance)
(2) 25 percent (one chance in four)
(3) 50 percent (two chances in four)
(4) 75 percent (three chances in four)
(5) 100 percent (certain)

23. Diagram B shows the gene combinations of a child who could be born to parents who each have the sickle-cell trait (*NS*). What is the chance that a child of this couple will have an *SS* gene combination, thus suffering from the disease of sickle-cell anemia?

(1) 0 percent (no chance)
(2) 25 percent (one chance in four)
(3) 50 percent (two chances in four)
(4) 75 percent (three chances in four)
(5) 100 percent (certain)

PRACTICE TEST

Question 24 is based on the following diagram and passage.

WAXING CRESCENT MOON

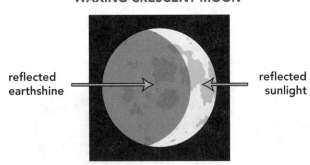

reflected earthshine →   ← reflected sunlight

When you see the Moon in its waxing crescent phase, the bright part of the face (the side facing Earth) occurs because of reflected sunlight. The rest of the face is much darker and is not reflecting direct sunlight. However, if you look carefully, you are often able to see the darker part of the face. The darker part is being weakly illuminated by *earthshine*—sunlight that travels to Earth, reflects off Earth, travels to the Moon, reflects off the Moon, and travels back to Earth.

24. During which phase of the Moon, would earthshine reflecting from the Moon be the least noticeable?

(1) waxing crescent

(2) quarter moon

(3) waxing gibbous

(4) full moon

(5) new moon

25. Iron (Fe) has an atomic number of 26 and an atomic mass of 56. Which of the following describes the nucleus of iron?

(1) 26 protons, 30 neutrons
(2) 26 protons, 56 neutrons
(3) 56 protons, 56 neutrons
(4) 56 protons, 82 neutrons
(5) 30 protons, 26 neutrons

Question 26 refers to the illustration below.

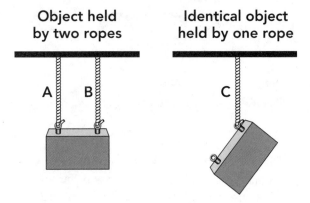

26. Which of the following statements is true?

(1) The load on rope C is half of the load on rope A.
(2) The load on rope C is the same as the load on rope A.
(3) The load on rope C is one and one-half times the load on rope A.
(4) The load on rope C is twice the load on rope A.
(5) The load on rope C is three times the load on rope A.

Question 27 refers to the following passage.

In a process called *fermentation*, yeast cells convert sugar into carbon dioxide and alcohol, releasing energy needed by the yeast for their own life processes. Yeast is used in the making of both bread and of alcoholic beverages.

27. For what purpose is yeast most likely used in the making of many types of bread?

   (1) Alcohol gives bread a better taste.
   (2) Alcohol evaporates as bread is baked.
   (3) Yeast removes excess sugar from bread dough.
   (4) The released energy is used to cook the bread.
   (5) Carbon dioxide causes bubbles to form in bread, giving bread its lightness.

Questions 28 and 29 refer to the following passage and illustration.

A *tide* is a periodic movement of ocean water that changes the level of the ocean's surface. High tides occur on the part of Earth nearest the Moon and on the part of Earth farthest from the Moon. Low ocean tides occur on each side of Earth between the positions of the high tides.

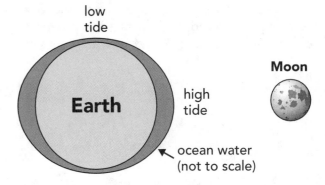

28. What can you infer causes the high tide that is on the side of Earth closest to the Moon?

   (1) the rotation of Earth
   (2) Earth's gravity
   (3) movement of the Moon in its orbit
   (4) the Moon's gravity
   (5) the rotation of the Moon

29. A spring tide, the highest (and lowest) of tides, occurs each time Earth, the Moon, and the Sun are approximately in a straight line. How many times each month does a spring tide occur?

   (1) one time
   (2) two times
   (3) three times
   (4) four times
   (5) more than four times

PRACTICE TEST

Questions 30 and 31 refer to the following passage and illustration.

As far back as 1100 B.C., ancient astronomers drew patterns of stars, called *constellations*, that they believed had special importance. They thought that stars were part of the surface of a large sphere in which Earth stood at the center. They pointed out that the stars in a constellation had the shape of a particular figure or design. *The Big Dipper* and *Orion* are well-known constellations seen in the Northern Hemisphere today.

Perhaps not surprisingly, different cultures have perceived many of the same stars of constellations in different ways. For example, in Greek culture the stars of Orion were seen as a hunter, while many of these same stars were seen in Japanese culture as a set of drums.

As science advanced, astronomers discovered that stars are at different distances from Earth. In fact, constellations are formed by stars that are not even near one another. Although scientific understanding of constellations has changed, star patterns still bring delight to all who view them.

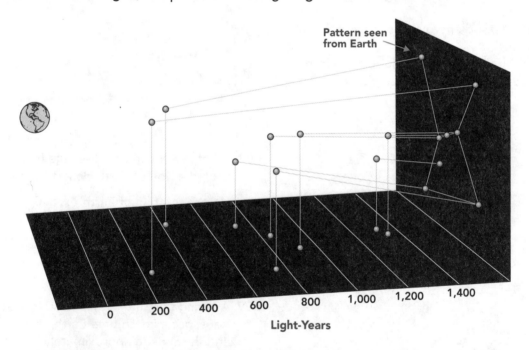

Seen from Earth, the stars in the constellation Orion appear near to each other. However, they actually are hundreds of light-years apart. (One light-year is a distance of approximately six trillion miles.)

30. How many stars in the constellation Orion are part of the group that make up the figure that in Japanese culture was said to be a pair of drums?

   (1) 2
   (2) 4
   (3) 8
   (4) 12
   (5) 16

31. In the group of stars shown above, about what distance is the star that is closest to Earth from the star that is farthest from Earth?

   (1) 250 light-years
   (2) 500 light-years
   (3) 1,100 light-years
   (4) 2,250 light-years
   (5) 3,800 light-years

PRACTICE TEST

Question 32 refers to the graph below.

WATER USAGE IN THE UNITED STATES

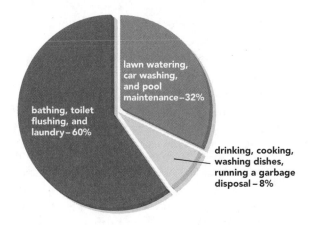

bathing, toilet flushing, and laundry – 60%

lawn watering, car washing, and pool maintenance – 32%

drinking, cooking, washing dishes, running a garbage disposal – 8%

32. The average family in the United States uses about 100 gallons of water each day. How much of this water is used for drinking?

(1) less than 8 gallons
(2) between 8 and 32 gallons
(3) between 32 and 40 gallons
(4) between 40 and 60 gallons
(5) more than 60 gallons

Question 33 refers to the following passage.

Earth's water cycle is the continuous movement of water between sources on the ground and in the atmosphere. The water cycle links all of Earth's solid, liquid, and gaseous water together. The energy that drives this cycle comes from the Sun and from Earth's interior.

33. Which of the following would <u>least</u> likely be considered part of Earth's water cycle?

(1) condensation
(2) ocean currents
(3) steaming geyser water
(4) evaporation
(5) snowfall

Question 34 refers to the following passage and diagram.

In the late 1600s Sir Isaac Newton explained why the Moon, even though pulled strongly by Earth's gravity, does not come crashing into Earth. The answer has to do with the Moon's motion.

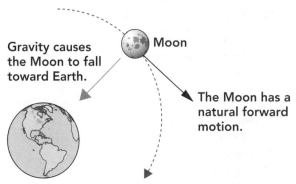

Gravity causes the Moon to fall toward Earth.

Moon

The Moon has a natural forward motion.

Sum of two motions = orbit

Newton explained that the Moon has two motions, both occurring at the same time.

• The first motion is the Moon falling toward Earth.

• The second motion is the Moon traveling forward in a straight line.

Newton concluded that the sum of these two motions (occurring at the same time) is a nearly circular path, the Moon's actual orbit around Earth.

34. What other motions could Newton have explained by using his ideas about the Moon?

A. the dropping of an apple from a tree
B. the movement of Earth around the Sun
C. the daily rotation of Earth

(1) A only
(2) B only
(3) C only
(4) Both A and B
(5) Both B and C

Questions 35–37 refer to the following passage and graph.

Fats are organic compounds made up of carbon, hydrogen, and oxygen. Fats provide the highest energy content of all nutrients and play an important role in many body processes.

- *Saturated fat*, found mainly in meats, dairy products, coconut oil, and palm oil

- *Unsaturated fat*, found in vegetable oils and in fish. Unsaturated fat can be either polyunsaturated or monounsaturated.

The American Heart Association recommends that people reduce the amount of saturated fat in their diets. Saturated fat contains *cholesterol,* a substance that the human body needs only in limited amounts. Excess cholesterol can lead to heart and blood-vessel diseases. Because the body naturally produces all the cholesterol it needs, any extra cholesterol added by food sources is potentially harmful to health.

Unsaturated fat, though, does not contain cholesterol, and, in limited amounts is good for you. *Polyunsaturated fat*—found in margarine, sunflower, soybean, safflower, and corn oils—supplies the body with needed fat. Better still is *monounsaturated fat*—found in olive, canola, and peanut oils. Monounsaturated fat also fills the body's needs for fat and has the advantage of actually decreasing blood cholesterol.

FAT CONTENT FOR SELECTED PRODUCTS
(present in one 25-gram serving)

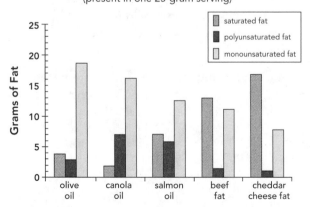

35. What is the main health risk that is associated with eating too much red meat and too many dairy products?

- (1) excessive blood clotting
- (2) an excess of body energy
- (3) excessive blood cholesterol
- (4) excessive weight gain
- (5) excessive weight loss

36. Which statement is supported by the graph?

- (1) Beef fat is lower in both saturated fat and monounsaturated fat than is canola oil.
- (2) Canola oil is higher in both saturated fat and polyunsaturated fat than is olive oil.
- (3) Salmon fat is lower in both saturated fat and monounsaturated fat than is beef fat.
- (4) Olive oil is higher in monounsaturated fat than any other listed product.
- (5) Cheddar cheese is lower in all three types of fat than is either salmon fat or olive oil.

37. Which products might be substituted in a diet for someone who is trying to lower cholesterol level by reducing consumption of saturated fat?

- A. olive oil for canola oil
- B. canola oil for olive oil
- C. beef for salmon
- D. salmon for beef

- (1) A and C
- (2) A and D
- (3) B and C
- (4) C and D
- (5) B, C, and D

PRACTICE TEST

Question 38 refers to the following passage.

Science has advanced to the point where population control is no longer a scientific problem needing to be solved. Scientific researchers have already provided a variety of birth-control methods that are now widely available in developed countries such as the United States. These same methods could readily become available in many developing countries where high birthrates often lead to overcrowding, mass starvation, and early death.

Unfortunately the lowering of birthrates in developing countries has more to do with political issues than with scientific ones. Studies indicate that birthrates are most easily brought under control in countries that grant democratic rights to all citizens. In these countries literacy rates increase and women are given economic status close to that of men. With increased literacy comes access to birth-control information. With democratic rights comes the right of women to make their own decisions regarding family planning.

38. Which of the following is the best summary of the passage above?

(1) For developing countries today, the most important factor in population control is democratic rights.
(2) For developing countries today, the most important factor in population control is birth-control methods.
(3) Effective birth-control procedures in developing countries remain a major scientific question.
(4) Developing countries have been more successful in family planning than have developed countries.
(5) Developing countries lack the scientific knowledge necessary to practice birth control.

Question 39 refers to the passage and graph below.

Petroleum use in industrialized countries is not proportional to the populations of those countries. For example, although the population of China is about four times that of the United States, the United States uses almost five times as much petroleum as China.

MAIN PETROLEUM CONSUMING COUNTRIES

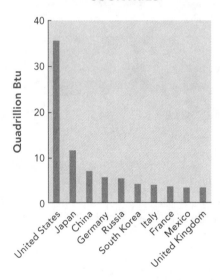

39. According to the information above, which of the following is most likely to happen during the twenty-first century?

(1) China will discover large reserves of petroleum within its own boundaries.
(2) China will further limit its use of petroleum.
(3) China will want access to more of the world's known petroleum reserves.
(4) The United States will decrease its use of petroleum so that China and other countries will have more petroleum available to them.
(5) China will reduce its population so that it will have less need for petroleum.

PRACTICE TEST

Questions 40–42 refer to the following passage and photographs.

A group of similar cells working together is called a *tissue*. The human body contains four main types of tissues:

Epithelial tissue

Nervous tissue

Muscle tissue

Connective tissue

- *Epithelial tissue*—cells that cover and protect underlying tissue. Examples are surface skin cells and cells that line the inner surfaces of body organs such as the lungs, stomach, and blood vessels.

- *Nervous tissue*—cells that carry nerve signals throughout the body. Examples are brain cells and cells found in sense organs.

- *Muscle tissue*—cells that can contract and relax for the purpose of moving skeletal parts or in the functioning of organs such as the heart and the stomach.

- *Connective tissue*—cells that join, support, cushion, and nourish organs. Types of connective tissue also form bones, cartilage, ligaments, and tendons.

40. What type of tissue is mainly involved in the sensation of taste?

    (1) epithelial
    (2) nervous
    (3) muscle
    (4) connective
    (5) individual brain cells

41. What type of tissue is the human body's first line of defense against airborne bacteria?

    (1) immune system
    (2) muscle
    (3) epithelial
    (4) connective
    (5) nervous

42. Cartilage is a type of fibrous connective tissue that is flexible and rubbery. Which part of the human body is made of cartilage?

    (1) biceps
    (2) ribs
    (3) fingernails
    (4) earlobes
    (5) brain

43. Which of the following represents the highest level of organization in an organism?

    (1) molecule
    (2) cell nucleus
    (3) cell
    (4) tissue
    (5) organ

PRACTICE TEST

Questions 44–46 refer to the following passage.

One of the most promising areas of research in human biology and medicine in the twenty-first century is stem-cell research. A *stem cell* is a living cell that has the potential to divide and give rise to a number of specialized cells. How stem cells do this is a mystery that scientists are now just beginning to understand.

What is known is that during the first few days following fertilization, a human embryo is made up entirely of stem cells.

- The human fertilized egg cell, called a *totipotent stem cell,* has the potential to divide and give rise to an entire human being, an identical twin of the original cell.

- Usually, a fertilized egg cell does not divide and produce more than one identical individual. Instead, the cell undergoes many cell divisions to become a sphere of cells called a *blastocyst*. The outer layer of blastocyst cells go on to form the placenta and other tissue needed by the fetus. The inner layer of blastocyst cells are pluripotent stem cells. The *pluripotent stem cells* specialize to become the cells of the fetus.

Researchers want to study pluripotent stem cells to possibly learn how to generate tissue and replacement organs that may be used to treat diseases. Also, an understanding of stem cells may help scientists determine why cells change and cause birth defects, cancer, and other devastating diseases.

While the potential value of stem-cell research is not disputed, the acquisition and use of human fetal stem cells is. The controversy concerns where and how researchers obtain fetal stem cells. One possible source is aborted fetuses. Opponents strongly object, claiming that this is immoral in itself and could also lead to the selling of aborted fetuses. This, in turn, could lead to pregnancies that have the goal of early abortion and the sale of the aborted fetus.

44. What is the best definition of *stem cell*?

    (1) A cell that has specialized to become a particular kind of cell such as a blood cell or a nerve cell.
    (2) A cell that is capable of forming tissue through repeated cell division and growth.
    (3) A cell that has the potential to divide and give rise to specialized body cells.
    (4) A cell that no longer has the potential to divide and give rise to specialized body cells.
    (5) A cell that is found in more than one type of body tissue.

45. Which of the following is <u>not</u> true about a pluripotent stem cell?

    (1) A pluripotent stem cell is part of the inner layer of cells in a blastocyst.
    (2) A pluripotent stem cell can specialize to become any one of a number of different types of cells.
    (3) A pluripotent stem cell develops from a totipotent stem cell.
    (4) A pluripotent stem cell can become a complete individual.
    (5) A pluripotent stem cell has a complete set of chromosomes.

46. What is the main source of controversy concerning human fetal stem-cell research?

    (1) the cost of human fetal stem cells
    (2) the length of time it takes to grow human fetal stem cells
    (3) uncertainty over the ownership of human fetal stem cells
    (4) how and where human fetal stem cells are obtained
    (5) what actually may be learned from human fetal stem-cell research

PRACTICE TEST

Questions 47 and 48 refer to the following passage and image.

A *hurricane* is a migratory tropical cyclone that is a yearly threat to islands such as Puerto Rico and the Virgin Islands in the Caribbean Sea and to coastal cities along the southeastern United States.

Hurricanes have high-speed winds that blow circularly around a low-pressure area known as the "eye." The diameter of the area of destructive winds may exceed 150 miles, with lesser winds extending out twice that distance. The speed of the circular winds ranges between 74 miles per hour (for category 1) to more than 155 miles per hour (for category 5). Along with strong winds, hurricanes are characterized by heavy rainfall and violent ocean waves.

Hurricanes move along a curving path at a speed of between 25 and 50 miles per hour—a speed much slower than that of their circulating winds. Just north of the equator, hurricanes tend to move in a northwesterly direction. As they more farther north, hurricanes begin to move more directly north or northeasterly.

When a hurricane strikes land, it can have devastating effects. In 1992 Hurricane Andrew struck Florida just south of Miami and left more than 40 people dead and 200,000 people homeless, and did an estimated $20 billion in property damage.

There is no known way to prevent, stop, or even slow a hurricane. However, the National Hurricane Center in Florida does try to issue early warnings of possible threatening hurricanes. The center collects data from radar, sea-based recording devices, and Earth-orbiting satellites to supply the public with up-to-the-minute information about each hurricane's progress. However, even with the most modern technology, the path of a hurricane cannot be predicted more than a few hours in advance. Hurricane prediction, like weather prediction, remains a very inexact science.

Hurricane season in the Caribbean lasts from early summer until well into fall. The arrival of October is especially welcome in the Virgin Islands where a special Thanksgiving Day is celebrated on October 25 to give thanks for the end of the hurricane season.

47. Suppose a hurricane is 400 miles off the east coast of the United States and is headed toward the shore of North Carolina. If a hurricane warning is now issued, about how much preparation time do people have before the hurricane may hit land?

    (1) between 2 and 6 hours
    (2) between 8 and 16 hours
    (3) between 2 days and 4 days
    (4) between 5 days and 7 days
    (5) between 1 week and 2 weeks

48. Suppose, following a hurricane warning, thousands of people evacuate their homes in a small town along the coast of North Carolina. The hurricane then turns north and misses the coast entirely. What may be one unfortunate result of this situation?

    (1) widespread damage to ocean life
    (2) the shutting down of the hurricane warning system
    (3) increased public support for the inexact science of hurricane prediction
    (4) decreased public confidence in the hurricane warning system
    (5) public anger that there is a hurricane warning system at all

PRACTICE TEST

Question 49 refers to the passage below.

Scientists in the nineteenth century believed that an invisible substance called *ether* filled all of space. They believed that ether was the substance that allowed light waves to travel from the Sun, the Moon, and stars to Earth, much like water is the substance through which water waves pass.

Then, in 1887 Albert Michelson, a German-born American physicist, did an experiment to measure the effect of ether on the speed of light. In the experiment two beams of light were sent in different directions over long distances. One beam was sent in the direction in which Earth moved in its orbit, while the other beam was sent perpendicular to the first beam. The beams were reflected by distant mirrors and returned to the starting point. The surprise discovery was that both beams traveled the equal distances in the same amount of time. Ether, if it existed, would have slowed one beam down relative to the other. This did not happen, and Michelson correctly concluded that ether did not exist in space.

The discovery that light travels freely through the vacuum of space was a major finding, and today the concept of ether is only an interesting footnote in the history of science. In 1907 Michelson received a Nobel Prize in physics for his work—the first American citizen to ever be awarded a Nobel Prize.

49. Nineteenth-century scientists compared light traveling through space with waves traveling through water. What is the most likely reason for this comparison?

    (1) Nineteenth-century scientists believed that light was a fluid much like water.
    (2) Nineteenth-century scientists believed that water vapor very likely filled space.
    (3) Nineteenth-century scientists did not have the knowledge that we have today.
    (4) Nineteenth-century scientists wisely used a familiar model as a starting point for the development of a new theory.
    (5) Nineteenth-century scientists did not have accurate tools of measurement.

Question 50 refers to the illustration below.

50. Why would an instrument similar to the one shown above be placed on the surface of the Moon?

    (1) to record temperature readings near an active volcano on the Moon
    (2) to record water vapor changes in the Moon's atmosphere
    (3) to record any movement of the Moon's surface
    (4) to record movement of surface dust on the Moon due to the solar wind
    (5) to record the number of asteroids passing near the Moon

Answers are on page 104.

Answer Key

1. (3) Because the shape of the pelvis is related to the function of the leg muscles, you can conclude that this shape is related to walking.

2. (4) A fallen dead tree would most likely be classified as biodegradable because it is decomposed by organisms, such as bacteria, fungi, termites, and ants.

3. (5) As shown in the drawing, magnetic field lines increase with electric current and increase near the wire itself.

4. (2) Microwaves are unable to leave the oven because they cannot pass through the reflective screen.

5. (3) The electrostatic force on the spheres is much greater than the gravitational force, and that is why the spheres are pushed apart.

6. (5) The best evidence will be obtained when the only difference between the diets is the amount of salt consumed.

7. (1) This is the only statement that cannot be checked by experiment. It is an opinion.

8. (5) This diagram shows how the Moon is illuminated by the Sun as the Moon orbits Earth, which accounts for the Moon's phases.

9. (3) Choices C and D are reasons why carbon dioxide gas makes an excellent fire extinguisher.

10. (4) When you use an inclined plane, you do the same total amount of work as when you directly lift an object. However, you use less force as you move an object, but you must move the object a longer distance.

11. (3) This molecule is the only one shown that has 5 carbon atoms and 12 hydrogen atoms, the same number as *n-pentane*.

12. (1) The decrease in pressure over the surface allows more dissolved gas to escape, which it does quickly.

13. (3) As ginger ale warms, the amount of dissolved gas decreases, causing the pressure above the liquid to increase.

14. (5) The apple is a nutritious part of many animals' diets and helps in the spreading of apple seeds.

15. (2) An animal that flies (such as a bird) obviously can disperse apple seeds over a much greater distance than an animal with any of the other listed characteristics.

16. (3) 30 percent = 9 percent of actually cultivated land + 21 percent of land potentially usable for agriculture.

17. (1) Iowa is a rich agricultural state, well known for its growth of corn.

18. (5) Desserts (too dry) and mountains (too mountainous) each make up about 20 percent of Earth's land.

19. (1) Each flip of the penny is independent of the other flips—whenever they occur. Each time you flip a penny, it is equally likely to land heads up or tails up.

20. (5) The cell is often called the *building block of life* or the *basic unit of life*.

21. (4) Jumping at the sound of a loud noise is a reaction of surprise, not homeostasis.

22. (3) Two out of four possibilities (50 percent) have an *NS* gene combination.

23. (2) One out of four possibilities (25 percent) has an *SS* gene combination.

24. (4) During a full moon, the amount of earthshine is very small compared to reflected sunlight.

25. (1) The atomic mass (56) refers to the sum of the protons and neutrons in the atomic nucleus. The atomic number (26) refers to the number of protons in the nucleus. (56 = 26 + 30)

PRACTICE TEST

26. (4) The load (weight) on rope C is twice the load on rope A. Rope C is holding the entire weight of the object, while Rope A is holding half the weight of the object, with Rope B holding the other half.

27. (5) Bread made without using yeast is called *unleavened bread* and is more dense and heavy than leavened bread.

28. (4) The Moon's gravity pulls ocean water toward the Moon, causing the ocean bulge on the side of the Moon.

29. (2) A spring tide occurs when the Moon is between Earth and the Sun, and when Earth is between the Moon and the Sun. Each of these positions occurs once a month.

30. (3) If you count the stars in the constellation, you will come up with the number 8.

31. (3) About 1,350 light-years − 250 light-years = 1,100 light-years.

32. (1) You can infer from the graph that it is less than 8 gallons.

33. (2) Ocean currents involve the flow of water within the ocean, but they do not involve the movement of water between the ground and the atmosphere.

34. (2) Newton's ideas about the Moon's orbit also explain the movement of Earth around the Sun as well as the movement of all other planets around the Sun.

35. (3) The consumption of excess animal fat is a known cause of an increase in blood cholesterol.

36. (4) Olive oil has the highest level of monounsaturated fat of any product listed on the graph.

37. (3) Both of these substitutions (B and D) will reduce a person's consumption of saturated fat.

38. (1) Science has provided methods of population control, but for political reasons, these methods are not available in all countries.

39. (3) As China develops technologically and the population begins to use more inventions that rely on petroleum, the country will demand an equal share of Earth's riches.

40. (2) Nervous tissue gives rise to all physical sensations, including taste and smell.

41. (3) Epithelial tissue lines the outside of your body as well as the inside passages of your respiratory and digestive systems.

42. (4) The rubbery part of both your earlobes and nose is made of cartilage.

43. (5) An organ contains each of the other answer choices as part of its structure.

44. (3) This is a definition is given in the first paragraph of the passage.

45. (4) A pluripotent stem cell exists only during the first few cell divisions, before the fetus develops.

46. (4) The fear is that human embryos will be produced for the sole purpose of obtaining fetal stem cells, resulting in the purposeful destruction of embryos.

47. (2) A hurricane moving 50 miles per hour would move a distance of 400 miles in 8 hours (400 ÷ 50); moving 25 miles per hour, the hurricane would take 16 hours (400 ÷ 25).

48. (4) Errors in the hurricane warning system often occur on the side of public safety, which can cause frustration for those who are evacuated.

49. (4) Starting with a familiar model is often an important first step in the development of any new theory.

50. (3) A seismograph, as shown, is used to record surface movements, such as earthquakes.

Evaluation Chart

On the following chart, circle the number of any item you answered incorrectly. Pay particular attention to areas where you missed half or more of the questions. For those questions that you missed, review the skill pages indicated.

Subject Area / Theme	Life Science (45%) (pages 157–254[s]; 459–496[c])	Physical Science (35%) (pages 255–336[s]; 533–577[c])	Earth and Space Science (20%) (pages 337–385[s]; 497–532[c])
Fundamental Understandings	14 questions 1, 2, 14, 15, 20, 21, 22, 32, 35, 39, 40, 41, 43, 44	10 questions 3, 4, 5, 10, 11, 12 13, 25, 26, 33	6 questions 16, 17, 18, 28, 29, 31
Unifying Concepts and Processes (pages 27–48[s])	1 question 42	2 questions 8, 19	0 questions
Science as Inquiry (pages 49–94[s])	2 questions 6, 7	1 question 9	1 question 24
Science and Technology (pages 95–114[s])	1 question 27	0 questions	1 question 49
Science in Personal and Social Perspectives (pages 115–138[s])	4 questions 23, 34, 36, 45	1 questions 50	2 questions 46, 47
History and Nature of Science (pages 139–155[s])	1 question 38	2 questions 37, 48	1 question 30

[s]*Contemporary's GED Science*
[c]*Contemporary's Complete GED*

Answer Key

Concepts and Processes in Science, pages 1–4

1. (1) Pea plants are annuals. They produce seeds and die at the end of the growing season.

2. (2) Carrots have a two-year life cycle, although they are normally eaten during their first year.

3. (4) A peony is a herbaceous perennial because the roots of this plant grow year after year. Only the aboveground part of the plant dies at the end of the growing season.

4. (3) Insulin is necessary for regulating blood sugar.

5. (2) The human body functions most efficiently with a controlled level of blood sugar (glucose).

6. (5) Homeostasis is a general term that refers to a body's tendency to maintain a stable set of internal conditions.

7. (1) A thermostat is also a regulating device. A thermostat regulates temperature.

8. (5) The total weight shown will not change because all the products produced by the combustion remain in the glass. The total amount of matter in the glass does not change even though the form of the matter does.

9. (3) On each flip the penny is equally likely to be heads up or tails up. The results of the previous flips do not affect the next flip.

10. (4) A molecule is made up of two or more atoms.

11. (2) There are 3 carbon atoms and 8 hydrogen atoms in the molecule: C_3H_8 is the correct way to write the molecular formula.

12. (1) The number of atoms remains constant even though reacting molecules are often replaced by different product molecules.

13. (4) The marbles are released (C), fall (A), bounce (D), and come to rest (B).

14. (1) The word *order* is used by scientists to describe the tendency of natural processes to be consistent and predictable.

15. (2) A sea horse is a fish, as are sharks and tuna.

16. (3) Human beings and reptiles are both vertebrates—organisms with a backbone enclosing a spinal cord.

Comprehending and Applying Science, pages 5–8

1. (4) Sunlight is called *white light*. White light is made up of a spectrum of colors as seen in a rainbow.

2. (2) On each side of this equation there are 3 carbon atoms, 8 hydrogen atoms, and 10 oxygen atoms.

3. (4) A deer and a blackbird are very different creatures with different diets and other needs.

4. (4) Scientists agree that vertebrates share a common ancestor. There is disagreement, however, on the exact characteristics of this common ancestor.

5. (4) Unlike an adult, a baby spends an equal amount of time in both light sleep and deep sleep.

6. (1) As the night progresses, the curve showing adult sleep is more and more above the line, into the light-sleep area.

7. (5) An adult spends the greatest amount of time in light sleep during the final hour, and dreaming only occurs in light sleep.

8. (2) The line for deer (the thin solid line) rises above normal at the beginning of the second 10-year period and dips below normal at the end of that period.

9. (3) Because the deer died of starvation, you can conclude that the problem was an inadequate food supply.

10. (4) Comparing the graphed lines, you can see that each of the three peaks for the cougar (the predator) follows a peak for the deer (the cougar's prey).

11. (3) Antibodies are passed from the mother to the infant through the mother's milk.

12. (5) Antibodies are produced in an animal and then injected into a person.

13. (2) When a person is exposed to a disease, antibodies are produced (shown as "primary immune response" on the graph). When the person is exposed to the disease a second time, the number of antibodies is much greater. Because of this, the person usually does not get the disease a second time.

Analyzing and Evaluating Science, pages 9–12

1. (2) The lamp may have stopped working because of a burned-out bulb or because the power has gone out. Maria's other thoughts don't seek to directly explain the problem.

2. (1) If the kitchen lights are still on, then they must not be on the same electric circuit as the lamp and the electric heater.

3. (2) The volume of the egg is determined by measuring the difference between the volume of the water plus the egg and the volume of the water alone.

4. (2) This fact is related to the reason that a sunset looks red/orange. At sunset the path of the sunlight goes through a great amount of atmosphere and the blue light is scattered out, leaving mainly red and orange.

5. (1) The boiling temperature of water is higher at the coast than in the mountains.

6. (4) A pressure cooker is used to increase water temperature and reduce cooking time in boiling water.

7. (5) A balance such as the one shown can be used to compare weight. The heavier object lifts the lighter object.

8. (1) Diffusion occurs because molecules of every substance are in rapid motion. This motion leads to a natural mixing of most, but not all, substances.

9. (4) The cider is holding all the sugar it can hold. Additional sugar settles on the bottom because the cider is saturated with sugar.

10. (5) The color of the cider will not change, because the amount of sugar in the cider will not change. Any more added sugar simply settles on the bottom of the glass.·

11. (2) By adding more apple cider you cause the solution to be no longer saturated. Because of this, sugar on the bottom of the glass will begin dissolving until the new solution is again saturated. If no sugar dissolves off the bottom, then you could conclude that sugar does not dissolve in apple cider.

12. (1) Rain and smog both absorb and scatter sunlight so that less of it reaches your eyes. Wind and temperature have no effect on light.

13. (4) In natural sunlight, paper appears bright white—the color of sunlight. Fluorescent light, on the other hand, contains a much higher percentage of yellow light than does sunlight, so in fluorescent light white paper often looks slightly yellow.

Science and Technology, pages 13–16

1. (5) The stirrer disperses microwaves evenly. If the stirrer isn't working correctly, food will cook unevenly.

2. (2) Most likely the pizza, which is in direct contact with the plate, warmed the plate as it heated up.

3. (1) Unlike a conventional oven, a microwave oven heats only food or other substances that contain water.

4. (3) If copper had the same resistance to electric current as does tungsten, the copper wire would also get hot and glow.

5. (3) Light energy is changed into electric energy, which then may be changed back into light energy.

6. (5) The term *solar-powered* is a clue that a photovoltaic cell is being used.

7. (1) Only laser light is coherent.

8. (4) The focusing property of a laser makes it useful as a long-distance signal carrier. None of the other examples make use of a focused beam of light.

9. (2) Lasers can be low energy, such as those used in eye surgery, or high energy, such as those used in welding. The energy of a laser depends on the laser's construction.

10. (5) The excretory system can be thought of as an output device for the human digestive system.

11. (4) The central processing unit (CPU) is the "thinking" part of a computer.

12. (3) Though not mentioned in the passage, lower cost is a major advantage of today's computers over earlier models.

Science in Personal and Social Perspectives, pages 17–20

1. (5) There are no sure cures for cancer, but much progress is being made toward that goal.

2. (3) The long-term effects of the new cancer treatments such as Gleevec are not known although early signs are hopeful.

3. (3) Nothing is mentioned in the passage about the cost of medicines being tried with Alzheimer's patients.

4. (4) This information is given in the first paragraph of the passage.

5. (5) This very serious condition is discussed in the third paragraph of the passage.

6. (3) Each of the other choices is a possible result of smoking during pregnancy.

7. (3) A fetus receives all of its oxygen from the mother's blood. Smoking decreases the amount of oxygen available in this blood.

8. (4) Sterilization is the only birth control method that is considered practically 100 percent effective.

9. (1) Female condoms are not a very effective birth control method, as the table shows.

10. (3) The best evidence would come from two groups of currently healthy people, one group that eats saccharin and one group that doesn't.

11. (2) Almost all dental work involves bleeding gums, so this choice is certainly not a practical approach.

12. (2) This statement is a brief and accurate summary of the passage.

13. (3) Cell walls give fruit, such as apples, its crunchy texture. A breakdown of cells walls results in a squishy texture.

14. (5) The misunderstanding of many people concerning food irradiation is not of direct interest to scientists trying to scientifically determine its safe use in food preservation.

History and Nature of Science, pages 21–24

1. (5) This choice, unlike each of the others, does not deal with an object's apparent weight changing because of submersion in water.

2. (4) The design of a submarine and the anatomy of fish are both related to their apparent weight under water. Each has a way of changing its submerged volume—the submarine by flooding air-filled compartments and the fish by inflating an internal air sac.

3. (3) This machine is known as Pascal's calculating machine. Pascal invented this machine so that his father would have an easier time doing math.

4. (2) Machines designed for warfare have a long history and continue to be the focus of much scientific research.

5. (2) Alchemists believed that all metals had the potential to become gold if only the right conditions could be set up.

6. (1) The fact that shined gold remains shiny while others metals tarnish gives gold special value even today.

7. (4) Franklin's experiment proved that lightning is a form of electricity—the flow of electrons.

8. (4) Edison was the father of many inventions, including the phonograph shown here.

9. (2) This type of research is occurring at the present time although the uncontrolled results depicted in the cartoon have not yet happened.

10. (3) The cartoonist may be saying, in part, "Just because scientists can do something doesn't mean that they should do it." This is a point of view that is shared by many people, including scientists themselves.

11. (3) Pythagoras saw great beauty in nature and in the relationship of numbers to nature. To him, the "music of the spheres" was real and had an almost religious significance.

12. (2) Pythagoras was one of the first scientists to recognize the importance of mathematics as a tool for understanding nature.

13. (4) This definition in given in the second paragraph of the passage.

14. (3) A photon does not have mass; or if it does, its mass is too small to measure accurately at the present time.

15. (1) This is a general property of wave motion but not of particle motion.

16. (3) This is a surprising fact of modern science. Often two (or more) models are considered correct even though they may seem to be inconsistent with one another.

Plant and Animal Science, pages 25–36

1. (4) The two highest peaks (absorption peaks) are for blue and orange.

2. (2) Mushrooms are white, not green, because they contain no chlorophyll. Plants that contain chlorophyll have green leaves.

3. (2) Light in the 500- to 600-nanometer wavelength range is green in color.

4. (3) There are about 20 growth rings seen on the cross section of the tree.
5. (1) Early death due to predators shortens the average life of animals in the wild.
6. (2) This conclusion follows from information given in the passage.
7. (1) Because the embryo is inside the yolk and both the yolk and embryo are within the shell, the yolk must get smaller as the embryo grows.
8. (4) Population-limiting factors cause a trout to die before its natural life span. All other choices except this one can end the life of a trout.
9. (2) With no population-limiting factors the trout population would grow, as shown by the dotted curve. The relatively constant population occurs because of population-limiting factors.
10. (3) At any one time the actual number of trout in a lake is slight different from its carrying capacity. The carrying capacity is the average number taken over a long period of time.
11. (1) Summer homes would most likely lead to increased water pollution and a decrease in the number of trout able to live in the lake.
12. (5) Moisture is present both in human lungs and on a piece of cheese.
13. (4) Both anaerobic bacteria and facultative anaerobes can live where there is limited oxygen, such as in the digestive tract of animals.
14. (4) There is no genetic diversity taking place in binary fission. Each daughter cell has exactly the same DNA as the parent cell.
15. (3) $100 \times 60 \times 50 \times 10 = 3,000,000$
16. (3) Keeping the planters on the same table would have little effect on plant growth. Each other choice plays an important role in the comparative growth of the plants.
17. (2) Planters A and C receive full days of sunshine but are given different amounts of water. These are the conditions being tested in this comparative study.
18. (3) None of the other answer choices can be inferred to be true from the information given in the graph.
19. (1) $10,000 - 6,000 = 4,000$

20. (5) The extinction of similar mammals in Europe would have no effect on the well-being of animals in North America.
21. (2) Both domesticated and wild horses are found throughout North America at the present time.
22. (2) This discovery was a result of the experiment.
23. (5) Though an interesting fact, it has nothing to do with the experiment.
24. (4) This is a prediction of something that may be discovered at some future time.
25. (1) This was the experiment itself.
26. (3) This is a hypothesis—a possible explanation of the findings of the experiment.
27. (2) Like a frog, a salamander is an amphibian, which develops lungs only during its water stage.
28. (4) Laying eggs and flying are characteristics of birds such as the killdeer.
29. (1) The sea horse is a remarkable fish that reminds an observer of both a horse and a kangaroo.
30. (1) This information is given in the final paragraph of the passage.
31. (3) A decreasing food supply is a threat to the long-term survival of orangutans.
32. (5) On each island the orangutan has adapted to conditions in a way that best ensures the survival of the orangutan species.
33. (2) Producing the greatest number of surviving offspring is characteristic of all life forms.
34. (4) Charles Darwin was the famous naturalist whose theory of evolution is a landmark in scientific thought.
35. (2) The gametophyte is known as the sexual generation of the fern.
36. (5) The spore does not result from the union of male and female egg cells. Because of this, the spore is considered to be an asexual reproductive cell.
37. (4) Because trees can be grown in a relatively short time, a forest is usually considered to be a renewable resource.
38. (1) Camouflage enables animals to hide from predators that might eat them.
39. (5) The waggle dance of bees is used to indicate both distance and direction.
40. (5) When they know the distance, bees can continually fly in widening circles at the approximate distance until they discover the flowers with nectar.

41. (5) The waggle dance is ineffective after sunset.
42. (2) Each of the other choices is a type of protection provided by a blood clot.
43. (2) Day 2 is the time shown between the 1 and 2 on the horizontal axis.
44. (3) Surrogate parenting results from progress made in the science of reproduction. Each other choice is a direct result of nature's design—processes which humans have found and have not changed.
45. (4) According to the Gaia hypothesis, Earth can recover from things that happen because of features of the Earth itself. Collision with an asteroid introduces something from outside Earth into Earth's system.

Human Biology, pages 37–48

1. (2) Choices (1) and (3) are not true. Choices (4) and (5) are true but are not the most complete summaries of the diagram.
2. (4) The occipital lobe is located just inside the back of the skull.
3. (2) Usually people who exercise enjoy the mild pain experienced from exercised muscles.
4. (5) Salted crackers are not a moist, high-protein food.
5. (2) Of the months listed July is the month when most people will barbecue at home and have picnics and other outdoor gatherings with friends.
6. (4) Of the choices given a campfire site is least likely to be a source of *Salmonella*.
7. (1) Hamburger buns are usually not moist and they are not a high-protein source.
8. (3) This is especially important for meats, salad dressings, mayonnaise, and any other meat or dairy products.
9. (3) This statement is a summary of the information provided on the chart.
10. (4) The need for vitamin pills by anyone who eats a nutritious diet is controversial. Some medical experts recommend vitamin pills, and some do not.
11. (3) For each of these injuries, bones are cracked or broken and a CT scan provides the clearest image of bone damage.
12. (4) This can be seen on the illustration.

13. (2) The soft spot on a baby's head is an area that is surrounded by bone that has not yet grown together.
14. (1) Extreme care must be taken with babies not to poke or strike the soft spot. There are only layers of skin and fluid that protect the brain below the soft spot.
15. (4) This is an illustration of the human heart, the organ that pumps blood throughout the body.
16. (2) This information is shown on the graph.
17. (4) This is the only statement that is supported by the information given in the passage and on the graph.
18. (2) Women's bodies differ from men's bodies due to the presence of mammary glands and female reproductive organs, all of which contain fat-rich tissue.
19. (3) Blood flows in one direction only, and this direction is controlled by valves.
20. (1) As shown in the third drawing, abnormal valve action occurs when a valve does not close properly and blood seeps back in the wrong direction.
21. (3) Lungs are used for breathing—taking in oxygen and expelling carbon dioxide.
22. (5) Many doctors now recommend that people who work outside or spend much time outside wear ultraviolet light-absorbing sunglasses.
23. (2) High-quality sunglasses are designed to absorb nearly all of the dangerous high-energy (UV) sunlight rays.
24. (4) For an object to be seen clearly, all light rays coming from it must all focus together on the retina.
25. (3) A convex lens decreases the focal length of light passing through the eye, ensuring that the light comes to focal point on the retina instead of behind it.
26. (2) This information is indicated on the graph.
27. (4) This statement is the most complete summary of information provided by the graph.
28. (1) You can infer from the graph that the oxygen consumption and the heart rate both reach steady-state rates and that these rates are determined by the rate of exercise. Exercise at a more strenuous level increases both heart rate and breathing rate.

29. (5) As seen on the graph, the jogger's heart is at a steady-state rate of oxygen consumption of about 1.75 liters per minute.

30. (3) Each jaw has 32 teeth.

31. (4) Although vocal sounds may also slightly depend upon jaw shape, the main differences relate to diet—what is eaten and how it is eaten.

32. (5) The smoker's heart would beat more rapidly in order to meet the body's oxygen needs. The smoker's blood is not able to carry oxygen efficiently.

33. (5) This is an interesting fact, but it is not directly related to alcohol and health.

34. (1) This statement is not true. Each of the other statements are facts that are shown in the illustration.

35. (2) The tip of the tongue is most sensitive to sweet tastes, particularly sugar or products that contain sugar.

36. (4) This is the only conclusion that can be drawn from looking at the illustration.

37. (1) This is a summary of information contained in the first two paragraphs of the passage.

38. (3) Epilepsy cannot be passed on from one person to another.

39. (5) The experiment is designed to test how quickly a person reacts to the dropping pencil.

40. (2) Alcohol interferes with both judgment and reaction time. For many people this would be a good test of sobriety.

41. (5) Each other choice is mentioned in the fourth paragraph as a symptom of heat exhaustion. Also mentioned is that a below-normal temperature is often a symptom.

42. (2) Lowering a victim's body temperature will help to prevent the serious consequences of heatstroke, such as brain damage and death.

43. (1) Each of the other devices directly affects the exercisers' access to cool air and water.

44. (4) Air temperature has nothing to do with the victim's immediate health needs.

Chemistry, pages 49–60

1. (4) When kept at a constant temperature, gas pressure increases as volume decreases.

2. (2) A steam engine uses gas pressure (steam) to move a piston up and down, thus changing the volume filled by the confined steam. None of the other choices deals with gas in a changing volume.

3. (2) Sulfur displays the properties of a nonmetal.

4. (3) Tellurium is shiny like a metal, but it is also brittle which is unlike a metal.

5. (4) Wind blowing over an iceberg can cause ice to become water vapor in the process of sublimation.

6. (2) Dew forms when water vapor in the air condenses on a cool surface to form drops of water. Condensation also occurs on a bathroom mirror when a shower is taken.

7. (4) There are 6 atoms of oxygen on each side of the reaction: $3 \times 2 = 6$ and $2 \times 3 = 6$.

8. (1) Freezing is a process of removing heat energy (molecular energy of motion) from an object, not a process of molecular penetration.

9. (3) The electrolysis of water results in the formation of hydrogen and oxygen gases.

10. (1) A higher-voltage battery would result in more electric current which would increase the rate at which electrolysis proceeds.

11. (3) Oil flows most slowly in the pipe with the largest diameter (C) and most rapidly in the pipe with the smallest diameter (B)

12. (4) This information is given in the first paragraph.

13. (1) Burning oil produces sulfur dioxide. None of the other choices lead to acid rain.

14. (2) Keeping warm is not protection against acid rain.

15. (5) Only in this choice are the number of iron atoms <u>and</u> oxygen atoms equal on both sides of the equation.

16. (3) After bonding, the chlorine atom has a negative (−) charge.

17. (3) The answer to this question is implied in the description of the property of equal pressure.

18. (5) The answer to this question is implied in the description of the property of effusion.

19. (1) The answer to this question is implied in the description of the property of loose molecular structure.

20. (4) The answer to this question is implied in the description of the property of diffusion.

21. (4) Strong molecular bonds are the reason that honey is so thick at room temperature.

22. (1) Cold engine oil is very thick until the engine starts and warms the oil.
23. (5) Cooling a liquid increases its viscosity. When frozen solid, a normal liquid no longer has viscosity.
24. (5) Both B and C are facts.
25. (1) Statement A is an opinion, most likely made from someone who does not find chemistry easy.
26. (2) This choice is the best summary of the information on the graph: that both vehicle emission of lead and air concentration of lead decreased at about the same rate (although one occurred after the other).
27. (4) Both of these facts must be known before a cause-and-effect relationship can be established.
28. (2) Heat is supplied to cookies that results in their baking.
29. (4) The reaction is exothermic because energy is given off. This also implies that water has less internal energy than the separated gases.
30. (1) The energy value of coal increases as the percentage of carbon increases.
31. (3) The economic value (and thus selling price) of coal depends on its energy content. The other choices are not directly related to the information on the table.
32. (4) Apple cider is a clear liquid composed mostly of water.
33. (2) A sail does not rely on or use gas in a closed container.
34. (2) The bulb lights only after salt is added to the water, indicating that dissolved salt turns water into a conductor of electricity.
35. (2) This equipment can test whether water contains salt, simply by seeing whether the bulb lights.
36. (3) The solution in a car battery conducts electricity, as does salt water.
37. (4) Because the line graph rises at first and then levels off, it is reasonable to conclude that both statements A and B are true. Statement C is also true, but it is not implied by the graph.
38. (4) The baking soda would partially neutralize the citric acid in the juice.

39. (5) Normal rain is slightly acidic with a pH value of 6. Acid rain is more acidic than normal rain and has a lower pH value.
40. (1) Lime helps neutralize the acidity of evergreen needles.
41. (3) Of the foods listed, only an egg has a pH value greater than 7.
42. (3) A quart and a liter (1,000 ml) hold about the same amount of liquid. The small beaker holds about one fourth of a quart, or about 250 milliliters (ml).
43. (5) This information is given in the first sentence of the passage.
44. (3) A candle requires heat (usually supplied by a lit match) to start it burning.
45. (1) The density of gold is more than twice the density of iron.
46. (1) Because ice is not as dense as water, ice floats on water. Ice is not as dense as water because the average distance between molecules is greater in ice.
47. (4) This is a scientific fact you should be aware of: the danger of both sulfuric acid and ozone.
48. (2) When a battery is recharged, the chemical reaction is reversed. The lead is returned to the two original compounds containing lead.
49. (3) The fact that food coloring stays on one side of the barrier is evidence of this fact.
50. (4) After a long period of time each side of the container will contain half of the water and half of the food coloring.

Physics, pages 61–72

1. (3) Heat is the energy of rapid molecular motion.
2. (5) The energy of the movement of machines of any type is called *mechanical energy*.
3. (2) Electromagnetic radiation includes invisible light as well as visible light.
4. (4) In a car battery chemical energy is converted to electrical energy.
5. (1) A superconductor carries electricity without experiencing the normal heat loss that normal conductors have.
6. (4) The items that cost the most to operate are also those that generate the most heat.
7. (4) You need to know how much time was spent on each of the two activities, but you don't need to know when the activities took place.

8. (2) Only this statement explains the fact that the magnets are being strongly pushed apart.

9. (4) A parachute operates on the principle of air resistance, not buoyancy.

10. (4) The gravitational force is greatest between the most massive objects at the closest distance.

11. (5) Sandpaper makes use of the principle of sliding friction and has nothing to do with the reflection property of light.

12. (4) Electrons in orbit move as far away from one another as they are able because of the repulsive electrostatic force.

13. (2) $75 \times 4 = 50 \times 6$

14. (3) This is the advantage of a hydraulic jack—to change a small amount of force into a great amount of force.

15. (2) As an example, suppose the area of the larger piston is 1; then the area of the smaller piston is 100 ($100 \times 30 = 3,000 \times 1$). A force of 30 pounds balances the weight of 3,000 pounds. Any force more than 30 pounds would lift the 3,000-pound car.

16. (4) The normal force applied to the brakes would not be applied because of the leaking fluid.

17. (3) This example demonstrates Pascal's law.

18. (3) The two forms of energy are equal at about the halfway point between 0 and 10.

19. (5) Each choice is true except choice (5).

20. (4) By wearing light-colored clothes and staying out of direct sunlight, you absorb the least amount of light energy.

21. (3) Of the examples given, this is the only one that relates electron motion to the creation of light.

22. (4) Of the objects listed, only a star gives off light that is created from processes not related to reflection or to electric current.

23. (1) The graph does not take air resistance into account. Air resistance affects the falling of a leaf.

24. (5) The data point (on the vertical axis) that indicates a speed of 80 feet per second is above the distance point (on the horizontal axis) of 100 feet.

25. (2) The point (on the horizontal axis) that indicates a distance of 50 feet is below the data point indicating a speed of about 56 feet per second.

26. (3) The faster an object falls, the more air resistance it feels.

27. (4) Diffraction is the spreading out of a wave around a barrier.

28. (3) Because different wavelengths (colors) of light refract by different amounts, a film of oil can split white sunlight into its rainbow of colors.

29. (2) An echo is an excellent example of the reflection of sound. An image seen in a mirror is an example of the reflection of light.

30. (5) Emilio bobs up and down because of the crests and troughs of the water waves. The distance between each two crests (or troughs) is one wavelength.

31. (1) It is only because light moves in straight lines that we are able to see shadows.

32. (2) Light changes direction as it moves from air to glass and then changes direction again as it moves from glass to air.

33. (2) The bicycle brake works because of friction: the friction of the brake pads against the wheel slows the bicycle.

34. (2) Static electricity is caused by separated charges that are not moving.

35. (5) Lightning starts as non-moving separated charges (static electricity) and ends as the separated charges rush together in a flash of light and a clap of thunder.

36. (3) Gears A and C turn in the same direction.

37. (3) A convection current is the movement of a warmer substance above a cooler one.

38. (4) The movement of electrons gives rise to a magnetic field, so it is reasonable to assume that it is also electron motion that gives rise to the magnetic field of a magnet.

39. (1) The fact that electric current does not add weight to a wire is not related to Ørsted's discovery relating electricity and magnetism.

40. (2) This is the only example of a force being felt by a current-carrying wire in a magnetic field.

41. (1) A microwave oven uses an external source of electricity (a wall socket) to produce electromagnetic energy. Each other item listed uses an electric generator to produce electricity.

42. (2) This information is given in the third paragraph of the passage.

43. (2) Drag (air resistance) slows an airplane. Thrust (push) is provided by jet engines.

44. (5) Only when the lift force is greater than an airplane's weight will an airplane gain altitude.

45. (3) Opening switch C breaks the circuit going through light 1 but does not affect the circuit for light 2.

46. (3) This tool is called a *caliper* and is designed to measure a small distance or a small thickness such as the thickness of a human hair.

47. (3) This information is given in the second paragraph of the passage.

48. (5) The word *calorie* is used today as a measure of food energy.

49. (4) Temperature is a measure of the energy of motion of atoms or molecules.

50. (2) When the two gases are at the same temperature, the average energy per atom or molecule is the same for both gases.

Earth Science, pages 73–82

1. (4) *Streak* refers to the identifying mark left behind when one material is rubbed against another.

2. (2) Luster refers to a material's ability to reflect light, or shine.

3. (1) The color of many minerals can be greatly changed by the addition of very small amounts of impurities (other substances).

4. (5) A diamond's commercial value as jewelry is very much determined by the shape of how it is cut along cleavage lines.

5. (1) Of the substances listed, only fluorite can be scratched by glass, which means it cannot be used to smooth a piece of glass.

6. (4) The hardness of glass must be between substances that can scratch glass (6 and above) and substances glass can scratch (5 and below).

7. (4) The prevailing winds blow more or less from west to east. These winds shorten flight time when flying from west to east.

8. (4) The shape of the Sun and the Moon are not directly related to the shape of Earth. Each of the other choices is true only because Earth has a spherical shape.

9. (4) The Mid-Atlantic Ridge is along the boundary line of two tectonic plates. The outpouring of magma and the presence of volcanoes along this ridge are evidence of this boundary.

10. (5) According to the illustration, the deepest part of the ocean is called the *trench*.

11. (2) At one time the continents sat next to each other and have slowly spread apart over the centuries. The ridge has most likely been midway between the continents during all this time.

12. (3) The temperature reaches almost $-60°F$ between the troposphere and stratosphere.

13. (3) This variation is shown on the graph between the altitudes of 20 and 50 miles.

14. (2) The plane stays within the troposphere where the temperature continually decreases with increasing altitude.

15. (1) Clouds are like a blanket that keeps Earth warm at night.

16. (4) Of the places listed, deserts are least likely to have nighttime clouds.

17. (3) A clear day allows the Sun to warm the land while a cloudy night traps the daytime heat.

18. (2) Container B contains only silt, and container C contains only sand.

19. (4) The soil should be dry and the depth of each sample should be the same. The weight of the soil is not directly related to its drainage property.

20. (2) According to the passage sand particles are larger than silt particles, which are larger than clay particles.

21. (4) This information is contained in the final paragraph of the passage.

22. (2) A mechanic would not have the need for a compass that each of the other people would have.

23. (5) Electric power lines produce strong magnetic fields that can affect the reading of a nearby compass.

24. (1) This is why the north pole of a compass is attracted to Earth's north magnetic pole—because it is actually the south pole of Earth's magnetic field.

25. (3) The lowest data point of the graphed line occurs at about 18,000 years ago.

26. (2) Only one cycle of cooling (steadily decreasing water level) and warming (steadily increasing water level) is represented by the graph.

27. (1) At the end of the ice age the weather began to warm up, ice began melting, and the ocean level began rising.

28. (4) The temperature is decreasing at a rate that is a little less than 4°C per 1,000 feet.
29. (3) If you extend the graphed line, you'll see that it crosses the dotted line at an altitude of about 8,000 feet.
30. (5) Erosion is the natural movement of rock. Gravity erosion is movement caused by gravity.
31. (5) Geysers, such as "Old Faithful" in Yellowstone National Park, are fairly predictable. Volcanoes are not predictable at all.
32. (1) Wind-blown sand moving close to the desert surface forms mushroom-shaped rocks in the desert.
33. (3) An anemometer is designed to measure wind speed.
34. (5) This is true according to the graph.
35. (2) At 20°C, air can hold a maximum of 15 grams per cubic meter of water vapor. At a relative humidity of 25 percent, air is holding 25 percent of its maximum value: 25 percent of 15 is a little less than 4 grams per cubic meter.

Space Science, pages 83–86

1. (2) The statement that shooting stars bring luck is just a fanciful opinion. Each of the other statements is an interesting fact.
2. (3) Only people who live near the equator have the opportunity of seeing all directions into the universe during the year.
3. (1) If a person were standing on the North Pole, Polaris would be directly overhead.
4. (4) Polaris stays in one position in the sky, with the other stars seeming to rotate around it as the night passes.
5. (2) To be in synchronous orbit, a satellite must revolve around Earth in the same amount of time in which Earth rotates on its axis.
6. (2) A satellite in synchronous orbit could not do either A or C.
7. (2) Hydroelectric power is electricity created by running water, which is not a potential source of power for astronauts.
8. (2) The flat panels on the Hubble Space Telescope are solar panels designed to provide power for the telescope.
9. (4) Earth's atmosphere is dirty with pollution, as well as with moisture, and interferes with the view of Earth-based telescopes.
10. (4) With no atmosphere the Moon has no wind. The Moon does experience each of the other choices, however.
11. (3) The Moon contains a smaller percentage of iron and other heavy elements than does Earth.
12. (1) The Moon's rotation period around its axis is the same as its revolution period around Earth. This is a very interesting situation, which means you can never see the other side of the Moon while standing on Earth.
13. (5) A black hole is an end result; it does not become something else.
14. (3) Because light gets trapped by a black hole, no reflected light or light produced at or near this object can escape from it. Thus, a black hole appears as a void in the sky.
15. (2) During a solar flare a stream of charged particles strikes Earth. Charged particles are most likely to disrupt radio communications.
16. (3) During a solar eclipse, the Moon goes directly between Earth and the Sun. The result is that the Moon blocks the main part of the Sun, and solar flares can be more easily seen shooting out from the Sun's surface.

Science Almanac

United States Customary Units of Measure

LENGTH
1 foot (ft) = 12 inches (in)
1 yard (yd) = 3 feet (ft)
1 mile (mi) = 1,760 yards (yds)

CAPACITY
1 pint (pt) = 16 fl. ounces (fl. oz)
1 quart (qt) = 2 pints (pt)
1 gallon (gal) = 4 quarts (qt)

WEIGHT
1 ounce (oz) = 16 drams
1 pound (lb) = 16 ounces (oz)
1 ton = 2,000 pounds (lbs)

Metric Units of Measure

LENGTH
1 centimeter = 10 millimeters
1 meter = 100 centimeters
1 kilometer = 1,000 meters

CAPACITY
1 milliliter = 1,000 microliters
1 liter = 1,000 milliliters
1 kiloliter = 1,000 liters

MASS
1 milligram = 1,000 micrograms
1 gram = 1,000 milligrams
1 kilogram = 1,000 grams

Scientific Units of Measure

UNIT	ABBREVIATION	MEASURES
ampere	amp	electric current
astronomical unit	AU	astronomical distance
calorie	cal	energy
hertz	Hz	frequency
joule	J	energy
kelvin	K	heat
ohm	Ω	electrical resistance
volt	V	electromotive force
watt	W	power

Classification of Organisms

HIERARCHY
Kingdom
Phylum
Class
Order
Family
Genus
Species

HUMAN CLASSIFICATION
Animalia
Chordata
Mammalia
Primates
Hominidae
Homo
Homo sapiens

OTHER EXAMPLES
Plantae, Fungi, Protista, Monera
Arthropoda, Mollusca, Platyhelminthes
Arachnida, Bivalvia, Cestoda
Scorpiones, Veneroida, Cyclophyllidea
Buthidae, Tricadnidae, Taeniidae
Centruroides, Tridacna, Taeniarhynchus
Centruroides vittatus, Tridacna gigas
(Striped Scorpion)   (Giant Clam)

Periodic Table of the Elements

Group: I   II

Period

Legend box:
- atomic number (6)
- number of electrons in each energy shell: 2, 4
- name: Carbon
- atomic mass (number of protons and neutrons): 12
- symbol: C

Period	Group I	Group II							
1	(1) **H** Hydrogen 1 — 1								
2	(3) **Li** Lithium 7 — 2,1	(4) **Be** Beryllium 9 — 2,2							
3	(11) **Na** Sodium 23 — 2,8,1	(12) **Mg** Magnesium 24 — 2,8,2							
4	(19) **K** Potassium 39 — 2,8,8,1	(20) **Ca** Calcium 40 — 2,8,8,2	(21) **Sc** Scandium 45 — 2,8,9,2	(22) **Ti** Titanium 48 — 2,8,10,2	(23) **V** Vanadium 51 — 2,8,11,2	(24) **Cr** Chromium 52 — 2,8,13,1	(25) **Mn** Manganese 55 — 2,8,13,2	(26) **Fe** Iron 56 — 2,8,14,2	(27) **Co** Cobalt 59 — 2,8,15,2
5	(37) **Rb** Rubidium 85 — 2,8,18,8,1	(38) **Sr** Strontium 88 — 2,8,18,8,2	(39) **Y** Yttrium 89 — 2,8,18,9,2	(40) **Zr** Zirconium 91 — 2,8,18,10,2	(41) **Nb** Niobium 93 — 2,8,18,12,1	(42) **Mo** Molybdenum 96 — 2,8,18,13,1	(43) **Tc*** Technetium 98 — 2,8,18,13,2	(44) **Ru** Ruthenium 101 — 2,8,18,15,1	(45) **Rh** Rhodium 103 — 2,8,18,16,1
6	(55) **Cs** Cesium 133 — 2,8,18,18,8,1	(56) **Ba** Barium 137 — 2,8,18,18,8,2	(57) to (71)	(72) **Hf** Hafnium 178 — 2,8,18,32,10,2	(73) **Ta** Tantalum 181 — 2,8,18,32,11,2	(74) **W** Tungsten 184 — 2,8,18,32,12,2	(75) **Re** Rhenium 186 — 2,8,18,32,13,2	(76) **Os** Osmium 190 — 2,8,18,32,14,2	(77) **Ir** Iridium 192 — 2,8,18,32,15,2
7	(87) **Fr** Francium 223 — 2,8,18,32,18,8,1	(88) **Ra** Radium 226 — 2,8,18,32,18,8,2	(89) to (103)	(104) **Rf*** Rutherfordium 261 — 2,8,18,32,32,10,2	(105) **Db*** Dubnium 262 — 2,8,18,32,32,11,2	(106) **Sg*** Seaborgium 263 — 2,8,18,32,32,12,2	(107) **Bh*** Bohrium 262 — 2,8,18,32,32,13,2	(108) **Hs*** Hassium 265 — 2,8,18,32,32,14,2	(109) **Mt*** Meitnerium 266 — 2,8,18,32,32,15,2

Rare Earth Elements

Lanthanide Series	(57) **La** Lanthanum 139 — 2,8,18,18,9,2	(58) **Ce** Cerium 140 — 2,8,20,18,8,2	(59) **Pr** Praseodymium 141 — 2,8,18,21,8,2	(60) **Nd** Neodymium 144 — 2,8,18,22,8,2	(61) **Pm*** Promethium 145 — 2,8,18,23,8,2	(62) **Sm** Samarium 150 — 2,8,18,24,8,2
Actinide Series	(89) **Ac** Actinium 227 — 2,8,18,32,18,9,2	(90) **Th** Thorium 232 — 2,8,18,32,18,10,2	(91) **Pa** Protactinium 231 — 2,8,18,32,20,9,2	(92) **U** Uranium 238 — 2,8,18,32,21,9,2	(93) **Np*** Neptunium 237 — 2,8,18,32,22,9,2	(94) **Pu*** Plutonium 244 — 2,8,18,32,24,8,2

III	IV	V	VI	VII	VIII

VIII

(2) 2
He
Helium
4

| | | | | | |

B (5) 2,3 — Boron 11
C (6) 2,4 — Carbon 12
N (7) 2,5 — Nitrogen 14
O (8) 2,6 — Oxygen 16
F (9) 2,7 — Fluorine 19
Ne (10) 2,8 — Neon 20

Al (13) 2,8,3 — Aluminum 27
Si (14) 2,8,4 — Silicon 28
P (15) 2,8,5 — Phosphorus 31
S (16) 2,8,6 — Sulfur 32
Cl (17) 2,8,7 — Chlorine 35
Ar (18) 2,8,8 — Argon 40

Ni (28) 2,8,16,2 — Nickel 59
Cu (29) 2,8,18,1 — Copper 64
Zn (30) 2,8,18,2 — Zinc 65
Ga (31) 2,8,18,3 — Gallium 70
Ge (32) 2,8,18,4 — Germanium 73
As (33) 2,8,18,5 — Arsenic 75
Se (34) 2,8,18,6 — Selenium 79
Br (35) 2,8,18,7 — Bromine 80
Kr (36) 2,8,18,8 — Krypton 84

Pd (46) 2,8,18,18,0 — Palladium 106
Ag (47) 2,8,18,18,1 — Silver 108
Cd (48) 2,8,18,18,2 — Cadmium 112
In (49) 2,8,18,18,3 — Indium 115
Sn (50) 2,8,18,18,4 — Tin 119
Sb (51) 2,8,18,18,5 — Antimony 122
Te (52) 2,8,18,18,6 — Tellurium 128
I (53) 2,8,18,18,7 — Iodine 127
Xe (54) 2,8,18,18,8 — Xenon 131

Pt (78) 2,8,18,32,17,1 — Platinum 195
Au (79) 2,8,18,32,18,1 — Gold 197
Hg (80) 2,8,18,32,18,2 — Mercury 201
Tl (81) 2,8,18,32,18,3 — Thallium 204
Pb (82) 2,8,18,32,18,4 — Lead 207
Bi (83) 2,8,18,32,18,5 — Bismuth 209
Po (84) 2,8,18,32,18,6 — Polonium 209
At (85) 2,8,18,32,18,7 — Astatine 210
Rn (86) 2,8,18,32,18,8 — Radon 222

Uun* (110) 2,8,18,32,32,17,1 — Ununnilium 269
Uuu* (111) 2,8,18,32,32,18,1 — Unununium 272
Uub* (112) 2,8,18,32,32,18,2 — Ununbium 277

* = Manmade

Eu (63) 2,8,18,25,8,2 — Europium 152
Gd (64) 2,8,18,25,9,2 — Gadolinium 157
Tb (65) 2,8,18,27,8,2 — Terbium 159
Dy (66) 2,8,18,28,8,2 — Dysprosium 163
Ho (67) 2,8,18,29,8,2 — Holmium 165
Er (68) 2,8,18,30,8,2 — Erbium 167
Tm (69) 2,8,18,31,8,2 — Thulium 169
Yb (70) 2,8,18,32,8,2 — Ytterbium 173
Lu (71) 2,8,18,32,9,2 — Lutetium 175

Am* (95) 2,8,18,32,25,8,2 — Americium 243
Cm* (96) 2,8,18,32,25,9,2 — Curium 247
Bk* (97) 2,8,18,32,26,9,2 — Berkelium 247
Cf* (98) 2,8,18,32,28,8,2 — Californium 251
Es* (99) 2,8,18,32,29,8,2 — Einsteinium 252
Fm* (100) 2,8,18,32,30,8,2 — Fermium 257
Md* (101) 2,8,18,32,31,8,2 — Mendelevium 258
No* (102) 2,8,18,32,32,8,2 — Nobelium 259
Lr* (103) 2,8,18,32,32,9,2 — Lawrencium 260

Planets in the Solar System

	DIAMETER (in miles)	DISTANCE FROM SUN (in AU)	LENGTH OF ORBITAL REVOLUTION (in Earth days)
Mercury	3,050	0.39	87.97
Venus	7,560	0.72	224.70
Earth	7,973	1.00	365.26
Mars	4,273	1.52	686.98
Jupiter	89,500	5.20	11.86
Saturn	75,000	9.54	29.46
Uranus	32,375	19.19	84.0
Neptune	30,937	30.06	164.8
Pluto	1,875	39.53	247.7

Web Links for Additional Instruction and Practice

The Biology Project

http://www.biology.arizona.edu/

This interactive online resource for learning biology contains tutorials and practice sets covering a range of topics, including biochemistry, cell biology, developmental biology, human biology, immunology, genetics, and molecular biology.

Biology ClassONline

http://www.mccsc.edu/~jracy/bcoindex.html

This site contains tutorials and quizzes on major topics in biology, including genetics, evolution, ecology, and plant and animal biology.

Basic Principles of Genetics

http://anthro.palomar.edu/mendel/

This site presents a tutorial on basic genetics. Instruction and practice quizzes are included.

Classification of Living Things

http://anthro.palomar.edu/animal/default.htm

This Web tutorial provides an introduction to basic biological taxonomy with an emphasis on the evolution of humans.

BodyQuest

http://library.thinkquest.org/10348/

This site lets you explore the many systems that make up the human body. Instruction, graphics, and practice games are included.

CHEMistry

http://library.thinkquest.org/3659/?tqskip=1

This virtual chemistry textbook serves as an interactive guide that allows you to expand your knowledge of basic chemistry.

Chemtutor

http://www.chemtutor.com

This site contains instruction in the fundamentals of chemistry.

Quia—Chemistry

http://www.quia.com/dir/chem/

This site contains games and quizzes related to topics in chemistry.

Learn Physics Today!

http://library.thinkquest.org/10796/

This site is designed to teach the fundamentals of physics, including mechanics, light and waves, and electricity.

Fizzics Fizzle!

http://library.thinkquest.org/16600/

This comprehensive guide to the world of physics is divided into three levels—Beginner, Intermediate, and Advanced.

zeroBio

http://www.execulink.com/~ekimmel/

This site contains a wealth of practice quizzes, games, and puzzles that deal with basic concepts in biology, chemistry, and physics.

Discovery Channel Online—Earth Journeys

http://www.discovery.com/exp/earthjourneys/earthjourneys.html

This site provides several interactive features related to the field of Earth science, including the opportunity to talk with Earth scientists and a virtual journey to the center of Earth.

Astronomy Picture of the Day

http://antwrp.gsfc.nasa.gov/apod/astropix.html

Each day this site features a different image or photograph of some aspect of the universe along with an explanation written by a professional astronomer.